# CURRENTS OF DEATH

Power Lines, Computer Terminals,
and the Attempt to Cover Up
Their Threat to Your Health

## PAUL BRODEUR

SIMON AND SCHUSTER
New York   London   Toronto   Sydney   Tokyo

Simon & Schuster
1230 Avenue of the Americas
New York, NY 10020

Simon & Schuster and colophon are registered trademarks
of Simon & Schuster, Inc.

Designed by Irving Perkins Associates
Manufactured in the United States of America

10 9 8 7 6 5 4 3 2 1

Library of Congress Cataloging-In-Publication Data

Brodeur, Paul
Currents of death: power lines, computer terminals, and the attempt to
cover up their threat to your health/Paul Brodeur.
p.  cm.
1. Electromagnetic fields-Health aspects.   2. ELF magnetic fields-
Health aspects.   3. Electric Lines-Health aspects.   4. Computer terminals
-Health aspects.   I. Title.
363.18'9-dc20                    RA569.3.B75    1989

ISBN 0-7432-1308-4

Part of this book appeared
originally in *The New Yorker*, in slightly
different form.

# Acknowledgments

The author wishes to thank Louis Slesin, the editor and publisher of *Microwave News* and *VDT News*, for his generous assistance and advice, which greatly facilitated the writing of this book. He also wishes to express appreciation to Charles Patrick Crow, Josselyn Simpson, Anne Mortimer-Maddox, Elizabeth Pearson-Griffiths, and Elizabeth Macklin—all of whom are members of the editorial staff of *The New Yorker*—for their valuable assistance in preparing the portion of the book that appeared in the magazine.

TO MILANE

# CONTENTS

# PART ONE

---

# Power Lines

# 1

---

# AN UNFUNDED STUDY

IN THE SPRING OF 1974, a woman named Nancy Wertheimer began to spend a day or two each week driving through residential neighborhoods in Denver, Colorado, stopping every once in a while in front of a house or an apartment building and climbing out of her car to have a look around. Each time she got back into the car, she jotted down in a small notebook whether the house or the apartment building was one story or more, whether it was constructed of wood or brick, and how close it was to a school, a factory, or a major thoroughfare. She then consulted a list of addresses and a map on the seat beside her before driving on to another dwelling.

If the residents of the neighborhoods she was visiting had taken notice of her, they would have observed a tall, somber-looking woman of forty-seven, with blue eyes, high cheekbones, and graying, curly hair, who was driving a ten-year-old Dodge Dart station wagon and acting as if she might be conducting a market survey to determine potential customers for, say, new siding, roofing, or automatic garage doors. In fact, Wertheimer was an epidemiologist, and she was just resuming her professional career after a six-year hiatus, and the list on the front seat of the station wagon, which she had obtained from the Colorado Department of Vital Statistics, contained the home addresses at birth of every child in the four-county Greater Denver area who had died of leukemia between 1950 and 1969, along with birth addresses of a matched list of children without cancer.

Wertheimer was born in New Haven, Connecticut, and was graduated from the University of Michigan in 1948, with a B.S. degree in psychology and biology. In 1954, she received a Ph.D. in experimental psychology from Harvard and Radcliffe, and a year later she moved with her husband, a psychologist, and their children to Boulder, Colorado, where she spent the next decade raising her family and working part time studying schizophrenia. In 1968, she was awarded a fellowship by the Department of Health, Education, and Welfare to investigate the relationship of childhood leukemia with maternal age and season of birth, but by then she had been divorced for several years, and she turned the fellowship down so that she could spend more time with her children, who were teenagers. When she finally got around to the leukemia study, it was with the idea that the disease might be linked to some kind of infection. For this reason, she was on the lookout for clustering among the addresses of its victims as she began driving through the residential sections of Denver in the spring of 1974. She had no funding for the study, so she was working on her own time and paying expenses out of her own pocket. "I was doing what epidemiologists sometimes do at the start of an investigation," she said recently. "I was looking for some kind of pattern."

As things turned out, Wertheimer did not come across any unusual clusters of childhood leukemia victims during her preliminary explorations. She did, however, notice something that seemed to be a bit out of the ordinary. "It was on my third or fourth trip," she recalls. "I had stopped at one of the birth addresses—an old woodframe house on the edge of the warehouse district in downtown Denver—and walked into an alley behind it, which ran between the backyards of two rows of similar houses. It was a mixed neighborhood—black, white, and Mexican—and many of the houses were run-down. Like most backyard alleys in the Denver area, this one was paved so that trucks could get through to pick up garbage and trash, and it contained a number of power poles strung with electrical wires and telephone lines. When I looked up I noticed an electrical transformer on a power pole behind the house I had stopped to examine. The transformer was black and cylindrical, and it was attached to the pole above a crossbar that supported several electrical wires, and as soon as I saw it I thought to myself, 'Hey, haven't I been seeing a lot of these things lately?' "

To begin with, Wertheimer did not attach any great importance to this observation. A short while later, however, she remembered seeing a magazine article accompanied by the photograph of a boy holding a pair of fluorescent lamp tubes that lit up because he was standing in the electrical field beneath a high-voltage power line. The article had suggested that the emanations from power lines might be hazardous, and with this in mind, Wertheimer contacted an official of the Public Service Company of Colorado and obtained the addresses of the dozen or so power substations in the Greater Denver area. At these installations, she learned, voltages as high as 230,000 volts (230 kilovolts, or kV), which were carried by high-voltage power lines from power-generating plants, were being stepped down by large transformers to 13,000 volts (13 kV), producing correspondingly higher currents for distribution along primary wires to neighborhoods within the city. (Electric current—a flow of charged particles that always produces an electromagnetic field—can be likened to water flowing in a pipe, and voltage can be thought of as the pressure that pushes current through a circuit.) Wertheimer now proceeded to map the locations of the substations in relation to the birth addresses of childhood leukemia victims and their matched controls, her idea being that if electric fields were somehow associated with the disease, the link might show up in the vicinity of high-voltage wires. The map showed only a faint correlation between the two, however, and she realized that she would have to test her initial observations in a more direct way.

Up to this point, Wertheimer had surveyed about fifty of some hundred and fifty dwellings on her list of addresses, so the next thing she did was revisit those dwellings, taking particular note of the presence and proximity of pole-mounted transformers. These transformers step down the 13,000 volts carried by the primary wires to the 240- and 120-volt levels required to operate electrical equipment and appliances in the average home. Current is then carried locally, for relatively short distances from the transformers, at these lower household voltages, in wires known as secondaries.

When Wertheimer began rechecking the birth addresses, she was not expecting to find anything unusual, but during the next few weeks an association between transformers and childhood leukemia kept leaping out at her. "It was baffling," she recalls. "The more I went around, the more the correlation cropped up. It simply wouldn't go away. Moreover, about the time I finished rechecking the old addresses and started visiting new ones, I began to observe an additional puzzling pattern. Houses in which young leukemia victims had lived were not just those closest to pole-mounted transformers, but, with unusual frequency, were also the houses next door—in other words, the second house from the transformer. The leukemia rate then dropped off sharply at the third house and was low in all subsequent houses on the line. I didn't know what to make of this, of course, but I was tremendously intrigued by it, even though I was sure it would turn out to be a fluke."

By autumn, Wertheimer was discussing her mysterious findings with a physicist friend named Ed Leeper, who lives in the foothills of the Rocky Mountains, just west of Boulder. He, too, was puzzled by her observations, of which he could make little sense, but since he knew that the various parts of an electrical distribution system carry uniform voltages, regardless of their distance from transformers, he reasoned that the phenomenon she was seeing could not be associated with the alternating electric field. (Current for industrial and household use is known as alternating current because it is generated and supplied at a frequency of 60 hertz [formerly called cycles per second], which means that it flows back and forth—first in one direction and then in the other—sixty times a second.) Leeper therefore suggested that if there were in fact a correlation, it would have to be with the magnetic field arising from the flow of current in the transformer, although he pointed out that such fields would drop off so rapidly that they should be negligible at the house. A correlation with magnetic fields would be especially interesting, however, because, unlike electric fields, which can easily be shielded by conducting materials, magnetic fields—invisible lines of force that interact with magnets and certain metals—readily penetrate almost anything that happens to stand in their way, including the human body.

In December 1974, Leeper fashioned a crude gaussmeter (a gauss is a unit of measure for magnetic-field strength) using a coil employed by television repairmen to demagnetize TV sets and an audio amplifier and speaker from an old walkie-talkie, and presented it to Wertheimer as a Christmas gift. The contrivance, which Wertheimer promptly dubbed "the gadget," provided a fairly accurate estimate of the strength of an alternating magnetic field by transforming it into an alternating voltage, which could be heard on the speaker as a hum. After learning how to operate it, Wertheimer drove to a middle-class neighborhood of well-kept woodframe houses on the northwest side of Boulder one morning in early January 1975, and parked her car near a pole-mounted transformer at the entrance to a backyard alley. "The poles in the alley were placed at the corner of every other lot, and they were strung in a manner that was typical of the power distribution system in the Greater Denver area," she remembers. "At the top of each pole was a thin primary wire carrying seventy-six hundred volts. Below that was a crossbar strung with several two-hundred-and-forty-volt secondaries. The thin primary had branched off an array of thicker primaries that came directly from a substation, and it had already supplied power to dozens of such transformers, which, in turn, provided reduced voltages that fed current into hundreds of homes from the two-hundred-and-forty-volt secondaries.

"When I got out of my car, I walked to the base of the transformer pole at the alley entrance and switched on Ed's meter, which gave off a loud hum that indicated the presence of a fairly strong magnetic field. I was prepared for that, of course, but when I started walking up the alley, away from the transformer pole, I found that contrary to what Ed and I had expected, the hum did not begin to diminish until I got past the next pole, at the far corner of the second house lot, from which several wires known as service drops carried current into nearby dwellings. These were the first service drops that reduced the current load fed into the secondary distribution line by the transformer.

"As I continued walking past the second pole, a strange thing happened—the hum indicating the presence of the magnetic field dropped off sharply. For a while, I couldn't understand the significance of this. Then it dawned on me that the point where the magnetic field fell off coincided with the pronounced decrease in the leukemia rate I had previously observed at the third house from the transformer. At that point, I talked with Ed Leeper and we realized that I had been mistaken in assuming that the increased incidence of leukemia I had seen at the two houses nearest the transformer was associated with closeness to the transformers. The association was really between those two houses and the first span of the secondary wire—the part that ran past the two houses on its way from the transformer pole to the second pole from which the initial service drops were made. This part of the wire was carrying current for all of the dozen or more homes being served by the secondary and that is why it was giving off the strong magnetic field I had heard as a loud hum. Once we made that connection—the link between the first-span secondary, with its relatively high current and strong magnetic field, and the two houses where so many leukemia victims had lived—I began to think I might be onto something.''

During the rest of 1975, Wertheimer spent a great deal of time making measurements of the magnetic fields surrounding first-span secondaries elsewhere in Greater Denver, and correlating those measurements with the birth addresses of children who had died of leukemia. The results of her preliminary analyses indicated that childhood leukemia victims had lived disproportionately often in dwellings that had first-span secondaries running past them. Toward the end of the year, Leeper built her a more sophisticated meter that could accurately measure 60-hertz magnetic fields in terms of gauss, and when she began using this instrument to assess the strength of the magnetic fields surrounding other types of wires in the electrical distribution system, she found that primary wires, which were used to carry high current from power substations, gave off relatively strong magnetic fields —as great as, or greater than, those of the secondaries she had been measuring. As a result, she predicted that if the magnetic fields produced by high current in secondary distribution lines were associated with excess leukemia among children, then high current in the primary wires would produce the same association.

Since Wertheimer formed this hypothesis on the basis of the birth addresses of childhood leukemia victims, sound epidemiological method dictated that it would be necessary for her to test the prediction with new data. For this reason, she went back to the Department of Vital Statistics and got the addresses at birth, as well as the addresses shortly before diagnosis, of all children in the Greater Denver area who had died of cancer of any kind between 1950 and 1973. Three hundred and forty-four such cases were included in the study. She also selected a matched-control population of 344 children, each of whom had a birthdate closely following that of a child who had died of cancer. She then devised a wiring configuration code that would allow her to classify each dwelling that had been lived in by a case or a control in terms of its proximity to high-current or low-current wiring.

High-current homes included those situated less than 130 feet from three-phase, large-gauge primary wires, or from an array of six or more small-gauge primary wires. The high-current category also included homes located less than 65 feet from an array of 3- to-5 small-gauge primary wires, and homes less than 50 feet from first-span secondary wires, which were defined as secondary wires that issued directly from a transformer and had not yet lost any current as a result of service drops. First-span wires serving no more than two single-family homes were considered to be in the low-current category, as were all other configurations of the distribution wires. Dwellings that were situated beyond the end pole of a secondary line, and thus had no distribution wires running past them, were considered the extreme example of low-current houses.

During 1976, Wertheimer visited the birth and diagnosis addresses of each of the cancer cases, the birth addresses of each of the controls, and the addresses at which control children had been living at the time their matched cases had been diagnosed with cancer. She then proceeded to draw a diagram describing the location, size, type, and proximity of the electrical wires and transformers she had observed in the vicinity of each of these homes. Once that was done, she analyzed the data and found that her prediction had held up: children who had lived in homes near high-current electrical wires had died of cancer at twice the rate seen in children living in dwellings near low-current wiring. The association was strongest among those children who had spent their entire lives in a high-current home. Particularly disturbing was the fact that of six children in the study population who had lived near high-current wires coming directly from power substations, all were cancer victims.

To satisfy herself that her findings were not the result of some artifact accidentally associated with high-current wiring, Wertheimer spent 1977 and the early part of 1978—still working on her own time and with her own money—analyzing and reanalyzing the data in relation to possible co-factors, such as population density and air pollution and noise from traffic congestion. When she could find nothing to explain her findings other than the association with the magnetic field resulting from high current, she and Leeper wrote a paper describing their study and its outcome, and submitted it to the *American Journal of Epidemiology*, which is published by the Johns Hopkins University School of Hygiene and Public Health in Baltimore, and is considered one of the foremost epidemiological journals in the world. After peer review, the paper was accepted and appeared in the *Journal*'s March 1979 issue, under the title "Electrical Wiring Configurations and Childhood Cancer." The first paragraph said:

> Electrical power came into use many years before environmental impact studies were common, and today our domestic power lines are taken for granted and generally assumed to be harmless. However, this assumption has never been adequately tested. Low-level harmful effects could be missed, yet they might be important for the population as a whole, since electric lines are so ubiquitous. In 1976–1977, we did a field study in the greater Denver area which suggested that, in fact, the homes of children who developed cancer were found unduly often near electric lines carrying high currents.

Wertheimer and Leeper went on to say that current flow (and, by extension, the magnetic fields induced by current flow) was always greatest in electrical wires leading from a distribution substation or a pole-mounted transformer. "At these points, the voltage has been stepped down and 'transformed' into current," they wrote, adding that "It was particularly homes close to these transforming points that were over-represented among our cancer cases." After acknowledging that magnetic fields can be canceled in ordinary wiring, where the return current tends to balance the supply current, they pointed out that such cancellation is imperfect in the vicinity of many dwellings "because the wires are often separated in space and, more importantly, because some of the return current does not flow through the wires at all, but returns instead through the ground, and particularly through the plumbing systems to which most urban electrical systems are grounded at each house." They explained that the ground current flows not only in street plumbing, but also through the pipes of houses, and thus produces magnetic fields in these homes which appeared to be related to the wiring configuration code near each house.

Aware that measurements had been made showing that household appliances could produce strong 60-hertz magnetic fields (an electric drill, for example, creates a field of 13 gauss at a distance of one centimeter), Wertheimer and Leeper warned that such measurements were not a valid index of exposure because they had been made very close to the appliance, and because appliance-generated fields usually fall off rapidly with distance. They pointed out that magnetic-field exposure to the whole body from normal use of household appliances rarely exceeds two milligauss (a milligauss is one thousandth of a gauss) for any extended period, while "ambient fields in a house due to nearby distribution wires or plumbing may sometimes reach those levels, or more, for hours or days at a time." This led them to conclude that if magnetic-field exposure was responsible for the increased incidence of childhood cancer they had observed, the duration of continuous exposure above some minimum threshold might be more important than the strength of the exposure per se.

To support the hypothesis that magnetic fields were associated with cancer, Wertheimer and Leeper said that upon analyzing a 1950 U.S. Public Health Service report correlating cause of death by occupation in men between the ages of twenty and sixty-five, they had found that workers who were frequently exposed to alternating-current magnetic fields—among them were power station operators, telephone linemen, power linemen, subway and elevated-railway motormen, electricians, and welders—had developed cancer at a significantly higher rate than the population as a whole.

Wertheimer and Leeper went on to suggest that alternating magnetic fields might indirectly affect the development of cancer by hindering the ability of the body's immune system to fight cancer. They also suggested that it is conceivable that the immune system might utilize electrical potentials occurring at cell surfaces, and they advanced the disturbing theory that when the human body was subjected to an electromagnetic background different from the level that had existed during man's evolution, the operation of the immune system might be altered. They then took pains to emphasize that whatever the reason for the correlation they had observed, the risk of cancer for children living in homes near high-current wiring was rarely increased by a factor of more than two to three.

"At the time our paper was published, I felt terribly ambivalent about our findings," Wertheimer recalls. "Part of the trouble stemmed from a deep concern that we might worry people unduly. I hated the idea of making people feel badly about their homes, especially if there was little or nothing they could do to correct the situation, and for this reason I was reluctant to speak to the press about our findings, or to make any special effort to publicize them. I should also say that the very idea that the kind of electrical wiring one could find in the streets of almost any town or city in America might be hazardous to health seemed— well, unbelievable. Additionally incredible was the fact that the cancer increase we had observed was associated with such weak sixty-hertz magnetic fields. The static magnetic field that is present at all times on earth was several hundred times as strong as those fields, and the magnetic fields that were described in the available medical literature as having essentially no biological effects whatsoever were thousands of times as strong. On the other hand, one out of every four or five families in our study lived in a dwelling that we had categorized as a high-current home, and that raised the possibility that, if the risk were real, alternating magnetic fields from electrical wiring could pose a very large public health hazard across the nation, as well as elsewhere in the world. Such a possibility obviously had to be investigated in greater detail, and for this reason I fully expected that the medical and scientific community would jump at the chance to study it."

# EARLY WARNING

WERTHEIMER'S EXPECTATION NOTWITHSTANDING, the medical and scientific community not only failed to consider her findings worthy of further consideration but dismissed them out of hand, on the ground that there was no experimental evidence to explain the hypothesis that the magnetic fields from ordinary high-current wires could cause cancer. Encouraged by such orthodoxy, the electric utilities industry tried to discredit her work, treating as heresy the idea that the nation's electrical distribution system might have adverse biological effects. As a result, more than eight years would elapse before the association she had discovered between high-current wires and cancer was corroborated in the United States. That belated confirmation might have been even further delayed had it not been for a courageous stand taken back in 1973 by Dr. Robert O. Becker, an orthopedic surgeon and research scientist at the Veterans Administration Hospital in Syracuse, New York, who had become convinced that the electromagnetic fields emanating from power lines could pose a serious threat to human health.

Becker is a bespectacled, blue-eyed, and slightly built man. Born in River Edge, New Jersey, in 1923, he went to Gettysburg

College, in Gettysburg, Pennsylvania, where he majored in biology. He got his medical degree from the New York University School of Medicine, in New York City, and subsequently became an orthopedic surgeon. In 1956, he was made chief of orthopedic surgery at the Veterans Administration Hospital, in Syracuse, where he soon became interested in the regenerative aspects of bone healing.

In 1958, Becker happened upon a paper written by a Soviet biophysicist who had measured negative electrical currents flowing from the wound sites of tomato plants whose branches had been cut off, and had found that these currents reversed their polarity from negative to positive as the wounds healed over and new branches began to form near the cuts. Becker was intrigued by this finding because he knew that the eighteenth-century Italian physician Luigi Galvani had observed an electrical current flowing at the wound site of an amputated frog leg, which later became known as the "current of injury."

In 1960, Becker conducted an experiment in which he measured the current of injury flowing at the wound of the amputated leg of a frog—an animal that lacks the capacity to regenerate its limbs—and compared it with the current of injury at the amputated leg of a salamander, whose remarkable regenerative power includes the ability to grow new limbs, tails, jaws, and eye lenses, as well as portions of its brain and heart. He found that while the direct-current electric potential was initially positive in both species, it became highly negative in the salamander as the salamander grew a new limb, and remained positive in the frog as its amputation wound healed over. In subsequent experiments, he demonstrated that the membranes of certain cells responded to the signal of the negative current by unlocking primitive genes within their nuclei, which proceeded to initiate the regeneration process. In order to explain why this regeneration was always appropriate—in other words, why the salamander replaced a foreleg with a foreleg and not with a hind leg—Becker went on to postulate that there might be an overall data-transmission and control system, which regulated growth and healing in the entire organism. He further reasoned that such a system might consist of a constant electric current associated with the nervous system, a current that could transmit information about injury to the brain and carry appropriate repair signals back to the site of the injury. After obtaining evidence for this theory by measuring DC electric current flowing throughout the body of the salamander, he hypothesized that the direction of flow and the magnitude of this current might also control brain activity and consciousness.

In 1962, Becker showed this to be the case when he induced deep anesthesia in a salamander by placing it in a strong magnetic field. By demonstrating that semi-conducting currents within the brain regulated brain activity, Becker affirmed the postulate of Albert Szent-Györgi, who had won the 1937 Nobel Prize in Medicine for isolating Vitamin C, and had theorized that cells might be semiconductors and thus act as pathways within the body for the passage of electric current. Becker also stimulated new interest in the biological effects of magnetic fields. The power of the lodestone to attract metal had been revered by the ancients, who believed that magnetism held the secret of life, but the biological effects of magnetism had remained largely outside the realm of accepted scientific inquiry in modern times. An exception was the work of Jacques-Arsène d'Arsonval, a French physician and physicist who was best known for demonstrating around 1890 that high-frequency electric currents could penetrate deep into the body and elevate the temperature of living tissue. In 1893, d'Arsonval discovered that a magnetic field oscillating at between 10 and 100 hertz could be perceived by humans as an intensely flashing light—a phenomenon that has since become known as magnetophosphene.

During the rest of the 1960s, Dr. Becker continued to experiment with electricity and living organisms, and contributed to the discovery that bone fractures in humans heal in the same electrically controlled manner that governs the regeneration of limbs in the salamander. (His work during this period is described in detail in a book entitled *The Body Electric: Electromagnetism and the Foundation of Life,* written with Gary Selden and published by William Morrow and Company, of New York City, in 1985.) Among his most significant achievements was his exploration of the influence exerted upon the activity of the brain by external magnetic forces, including those of the magnetic field of the earth. In 1963, he and Howard Friedman, a psychologist at the VA Hospital in Syracuse, showed that there was a relationship between the admission of patients to the psychiatric services of hospitals and the occurrence of solar magnetic storms. In 1965, they exposed human volunteers to pulsed magnetic fields of similar frequency and considerably greater strength than those associated with magnetic storms, and found that this significantly slowed the volunteers' ability to react to the appearance of a light by pressing and releasing a telegraph key. A year earlier, however, a Soviet investigator named Yuri Alexandrei Kholodov had reported that exposure to even stronger magnetic fields caused areas of cell death in the brains of rabbits. When Becker and Friedman learned of this, they discontinued their experiments with humans and advised other researchers to do the same. They then replicated Kholodov's experiment and discovered that the magnetic field produced stress in the rabbits, which activated a preexisting but quiescent brain disease that produced the lesions Kholodov had observed.

The fact that magnetic fields could induce stress worried Becker because of an earlier finding by the brilliant endocrinologist Hans Selye that prolonged stress can adversely affect the immune system. (It had already occurred to Becker during the course of his experiments with salamanders that if electric current could govern benign growth, it might also play a role in the development of malignant growth.) He became even more worried when Friedman exposed chimpanzees to pulsed magnetic fields of much lower strength than those employed by Kholodov, and found that the animals excreted elevated levels of adrenal cortical hormones—a well-recognized sign of stress. He was particularly concerned by the fact that the biologically active magnetic fields that he and Friedman had used in their experiments lay within the extra-low-frequency (ELF) range of one to thirty hertz—the principal frequency range of the earth's magnetic field, to which the human body had been tuned during its entire evolutionary period —because this suggested that man-made ELF fields, such as those created by the 60-hertz electrical distribution system, whose emanations permeate the modern environment, might have serious consequences for human health.

In the summer of 1972, Dr. Becker warned a meeting of the Institute of Electrical and Electronic Engineers about "the continuous exposure of the entire North American population to an electromagnetic environment in which is present the possibility of inducing currents or voltages comparable with those now known to exist in biological control systems." At the same time, he called for an early program to study the problem of human exposure to electromagnetic energy. A year later, he was invited to serve on a seven-man advisory committee that was being convened by the Navy's Bureau of Medicine and Surgery to review a research program the Navy had instituted on the biological and ecological effects of ELF radiation. The committee met at the Naval Medical Research Institute in Washington, D.C., on December 6 and 7, 1973, and Becker's subsequent revelation of its deliberations marked a dramatic turning point in his career, initiating a bitter controversy over the safety of 60-hertz electric and magnetic fields which continues to this day.

Navy officials had become interested in ELF radiation back in 1958, when they learned that radio waves oscillating just above the 60-hertz range could penetrate seawater sufficiently to provide communication with deeply submerged submarines. Because the wavelength of such a signal is nearly 2,500 miles, it was feared at the time that ELF transmitting antennas would have to be unduly large. This problem was solved, however, by Nicholas Christofilos, a brilliant Greek-born physicist working for the Department of Defense, who suggested that a portion of the earth's interior could be used as a launching pad to propagate ELF signals. During the early 1960s Christofilos's concept was successfully tested, and in 1969 the Navy and the RCA Corporation built an ELF test facility near Clam Lake, Wisconsin, by burying twenty-eight miles of insulated cable in the low-conductivity granite bedrock of the Chequamegon National Forest.

Soon thereafter, the Navy proposed to construct a 22,500-mile-square antenna system—to be called Project Sanguine—by burying 6,000 miles of cable in bedrock elsewhere in northern Wisconsin and in the Upper Peninsula of Michigan. The idea was to form a giant grid so that electric current generated by transmitters would pass through the antenna cables and flow deep into the earth along the bedrock, creating a global ELF radio field extending up to the ionosphere—a region of electrons and electrically charged particles in the upper atmosphere between 40 and 250 miles above the earth—which would reflect a portion of the ELF field into the world's seas and oceans.

Environmental groups mounted opposition to Project Sanguine soon after the Clam Lake facility was constructed, because some people living near it reported that they were receiving electrical shocks whenever they turned on their water faucets, and because ordinary wire fences suddenly became electrically charged. At that time, the environmentalists claimed that the alternating mag-

netic fields generated by the ELF antennas could produce voltages in nearby electrical conductors, such as power lines, telephone lines, and wire fences, and that this posed a potential
health hazard to the several hundred thousand people who would
be living within the boundaries of the Sanguine system.

From the beginning, the Navy tried to claim that there were no
harmful biological effects caused by ELF electric and magnetic
fields, and that the environmental effects of Project Sanguine,
which was designed to operate at a frequency of about 76 hertz,
would be no different from the effects caused by 60-hertz power
lines. As for any magnetic-field hazard that might exist, Navy
officials sidestepped this by emphasizing in a 1972 environmental
impact statement for Project Sanguine that many common household appliances—among them hair dryers, can openers, and food
mixers—give off strong magnetic fields. They made no mention
of the fact that these fields fall off rapidly within a few inches of
the appliances, and thus could not be compared with the magnetic
fields that would be given off by the large antennas of Sanguine
or, indeed, by power lines. Such omissions notwithstanding,
studies financed by the Navy's own ELF biological research program had by 1973 begun to cast doubt upon claims that Sanguine
was safe, and for this reason the advisory committee on which
Becker had been invited to serve was formed.

Becker has said that the only thing sanguine about Project Sanguine was its name. In any event, he and the other members of
the committee found little to be cheerful about when they met to
evaluate the results of the Navy's ELF research. To begin with,
they learned that an experiment in progress at the Naval Aerospace Medical Research Laboratory in Pensacola, Florida, had
produced significant rises in serum triglycerides—a warning of
possible stress—in the blood of nine out of ten human volunteers
who had been exposed to a low-intensity magnetic field. They
also learned that abnormally high triglyceride levels had been
measured in six out of eight workers at the Navy's Sanguine test
facility at Clam Lake. There were other disturbing findings. In a
study conducted at the Brain Research Institute of the University
of California at Los Angeles, monkeys exposed to ELF fields
showed a decrease in their ability to perform lever-pressing tasks
—an indication that ELF might produce adverse behavioral effects—and in a study conducted by scientists at the Naval Air
Development Center in Johnsville, Pennsylvania, rats exposed to

weak ELF fields did not gain weight as readily as unexposed control animals. Still other experiments showed that long-term exposure to ELF retarded the growth of chicken embryos, and that alternating-current magnetic fields produced disorientation in ring-billed gull chicks, which suggested that magnetic fields could affect bird migration.

Not surprisingly, the committee reacted strongly to these findings. Its members recommended as "urgent and absolutely necessary" that further studies be made of the effects of the planned Sanguine ELF fields on triglyceride levels in humans and animals, and that extensive investigations be made of the influence of ELF on the central nervous system, on cells and chicken eggs, on the growth rate of rats, and on bird migration and orientation. "Since the immune system is such an important and critical defense mechanism," the committee advised the Navy, "it behooves Project Sanguine to investigate the operational efficiency of immune mechanisms during and after exposure to ELF electric and magnetic fields." Most important of all from the public health point of view, the committee voted unanimously to go on record as recommending that the Electromagnetic Radiation Management Advisory Council (ERMAC)—a nine-member panel that had been established in 1968 by the President's Office of Telecommunications Policy to advise the President on the biological hazards of microwave and radio-frequency radiation—"be apprised of the positive findings evaluated by this Committee and their possible significance, should they be validated by further studies, to the large population at risk in the United States who are exposed to 60 hz fields from power lines and other 60 hz sources."

For its part, the Navy effectively suppressed the findings and recommendations of the advisory committee by compiling them in a thirty-one-page report marked "For Official Use Only," and limiting its distribution to members of the committee and other people who had participated in the two-day meeting. As a result, the warning that the committee had voted to send to ERMAC concerning the safety of the nation's 60-hertz electrical distribution system went undelivered. Indeed, the whole matter might have been swept under the rug if Dr. Becker had not happened to read a copy of the Lowville *Journal and Republican* upon his return from Washington. Lowville is a small town on the western slope of the Adirondacks, about 60 miles northeast of Syracuse, where he and his wife had recently bought land for a vacation retirement home, and the *Journal and Republican* contained a notice about a proposal of the Power Authority of the State of New York (PASNY) to build several 765,000-volt (765 kV) power lines in the state, in order to import electricity from a giant hydroelectric power-generating station that was under construction at James Bay in northern Quebec. According to the notice, the first of the proposed 765,000-volt lines would run from Massena, a city on the St. Lawrence River at the Canadian border, to the town of Marcy, 10 miles northwest of Utica—a total distance of about 155 miles—and would pass through the outskirts of Lowville on its way.

Three days later, Becker wrote a letter to Henry Diamond, the state Commissioner of Environmental Conservation, with a copy to Joseph Swidler, chairman of the New York State Public Service Commission (PSC). After giving a brief description of Project Sanguine, Becker told the officials that its antennas would transmit electromagnetic radiation at frequencies similar to those of 60-hertz power lines, and at electric- and magnetic-field strengths "*lower* than those that would be present along the proposed 765 kV line and for some distance out from the line." Becker informed them that the Navy's advisory committee had concluded that the civilian population might be at risk from exposure to the stronger electromagnetic fields emanating from 60-hertz power lines. He urged that permission for the construction of the Massena-to-Marcy line be withheld until the Project Sanguine experiments were completed and an evaluation of the true extent of the hazard could be made. By way of furnishing corroboration, he advised the two men to ask the Navy's Bureau of Medicine and Surgery for information regarding the advisory committee's findings and recommendations.

Chapter

# SPEAKING OUT

AT THE TIME DR. BECKER WROTE HIS LETTER, public hearings for PASNY's proposed Massena-to-Marcy power line were under way in Massena, and within a few months hearings began near Rochester for a 66-mile-long, 765,000-volt transmission line that the Niagara Mohawk Power Corporation and the Rochester Gas & Electric Company wanted to build from Rochester to Oswego. Experts for PASNY and the two power companies testified that there would be no biological hazard from the electromagnetic fields of the proposed transmission lines, but when a citizens' group raised the health issue at the Rochester hearing, the Public Service Commission felt obliged to look into the matter. In July 1974, a twenty-seven-year-old lawyer from the PSC named Robert Simpson visited Becker at his office at the VA Hospital in Syracuse, and listened intently as Becker described in detail the adverse biological effects of ELF fields that had been turned up by the Navy's research program on Sanguine. (The PSC had requested a copy of the advisory committee's report on the program, but the request had been denied by the Navy's Bureau of Medicine and Surgery, on the grounds that the information it contained was classified.) Also present at the meeting in Becker's office was Andrew Marino, a thirty-three-year-old biophysicist with a law degree, who had been working in Becker's laboratory for nearly ten years. Marino told Simpson about a study he had recently conducted in which rats exposed to 60-hertz electrical fields gained less weight and drank less water than unexposed control animals. Rats in a related study showed altered levels of blood proteins and enzymes.

At the conclusion of the meeting, Simpson asked Becker and Marino to testify about the biological effects of ELF radiation at the Rochester hearing; the two men not only agreed but decided to testify without remuneration. In October, they sent separate reports to Simpson, who forwarded them to Niagara Mohawk and Rochester Gas & Electric. In addition to describing the results of his rat experiments, Marino cited eight studies that had been published in the medical literature, which showed that ELF radiation caused biological effects in humans or animals. For his part, Becker wrote that ELF fields were biological stressors and testified that as a physician he would have to assume that their effects could be harmful. The suggestion by scientists that electromagnetic energy from power lines might be hazardous to human health appeared to take Niagara Mohawk and Rochester Gas & Electric by surprise. The two companies requested that the Rochester hearing be postponed for a year, so that they could call additional expert witnesses and produce new testimony.

By the time the PSC hearings resumed in late 1975, state officials had decided that the application of Niagara Mohawk and Rochester Gas & Electric for the 765,000-volt transmission line from Rochester to Oswego would be joined with PASNY's application for the Massena-to-Marcy line, and that the testimony of Becker and Marino would be presented in behalf of the PSC's staff, who are specifically charged by law with the responsibility of representing the interests of the public. In February 1976, however, the PSC bowed to pressure from Governor Hugh Carey and authorized construction of the Massena-to-Marcy line. (The governor had sent a bill to the New York State Legislature that, if passed, would have ordered the PSC to approve PASNY's power-line project.) Thus, the issues that remained to be decided by the new hearings were whether the right-of-way of the Marcy-to-Massena line, which PASNY and the utilities wanted to be 250 feet, would be expanded as a result of health and safety considerations, and whether the application of Niagara Mohawk and

Rochester Gas & Electric to build any more 765,000-volt lines in
the state would be approved.

During the year-long postponement, Becker and Marino had
combed the medical literature and found more than twenty additional studies that described biological effects caused by exposure
to ELF electromagnetic fields in humans and animals. Meanwhile, Marino had conducted repeated experiments in which he
exposed rats to 60-hertz electric fields whose strength was almost
identical to what could be expected at ground level beneath a
765,000-volt line, and found changes in body weight and blood
chemistry in the animals. He then subjected three generations of
mice to the same fields—exposing the animals from birth through
maturity and mating—and found that the second and third generations of exposed animals were severely stunted when compared to unexposed control animals. He had also found a
significantly increased mortality rate in the exposed animals.

After the hearings resumed, they dragged on for nearly two and
a half years, producing a record of more than fourteen thousand
pages of testimony by thirty-one witnesses. (The story of these
hearings has been recounted in a book entitled "The Electric
Wilderness," which was written by Andrew Marino and Joel
Ray, a freelance writer living in Ithaca, and published by the San
Francisco Press in 1986.) The case against the transmission lines
rested chiefly on Marino's experiments and on some thirty-two
other studies showing that ELF fields could cause biological effects, such as disturbances in human biorhythms, slowed reaction
time in monkeys, disorientation in birds, and increased heartbeat
rate in fish. Equally important was the testimony of Becker—the
only medical doctor to appear as a witness for either side. He
stated that the weight of this evidence suggested that biological
effects would occur in human beings who were subjected to long-
term exposure to electromagnetic fields from the lines. Both
Becker and Marino took the position that any biological effect
should be considered hazardous until it had been shown clearly
and convincingly to be harmless, and they urged that an extensive
program of research be undertaken to determine the potential
health risks of power-line radiation. They also declared that to
expose people to power-line fields without their permission
amounted to human experimentation without informed consent.

To counter their testimony, Niagara Mohawk and Rochester
Gas & Electric retained the services of several experts. Chief

among them was Herman P. Schwan, a professor of electrical engineering at the University of Pennsylvania's Moore School of Electrical Engineering in Philadelphia. Schwan, who had previously been a professor at the Kaiser Wilhelm Institute of Biophysics in Frankfurt, Germany, had come to the United States in 1947 under the auspices of Operation Paperclip—a government program designed to make the talents of German scientists available to the nation's defense effort. He soon began to conduct research on the biological effects of microwave radiation for the U.S. Navy, which was seeking to develop an exposure standard for personnel who were operating high-power radar, and in 1953, he proposed to the Navy that a safe level of exposure for humans could be set at ten milliwatts of power per square centimeter of body surface. (The power, or intensity, of microwaves, radio waves, and other high-frequency non-ionizing electromagnetic radiation is customarily expressed in terms of power density, which is the amount of radiation that flows each second through a square measure of space.) Schwan arrived at the ten-milliwatt level on theoretical grounds, basing it largely on the fact that energy from the metabolism of food is normally dissipated by the surface of a human body at rest at the rate of about five milliwatts per square centimeter. His assumption was that an additional heat load of another five milliwatts per square centimeter could be applied by external forces, such as microwave radiation, without causing a significant rise in the body's temperature. Although the ten-milliwatt standard was a thousand times the level of occupational microwave exposure then considered to be safe in the Soviet Union, it was adopted as a tentative standard in the late 1950s by the U.S. Army, the Navy, the Air Force, the Bell Telephone Laboratories, and the General Electric Company. By 1966, it was generally accepted as a guideline for occupational exposure by the electronics industry and by most state and federal health agencies.

When Schwan testified at the PSC hearings, he said that the power levels of the electromagnetic fields emanating from the proposed 765,000-volt transmission line could not interact with the body's cells to produce internal fields capable of heating tissue or exciting nerves, and that the risk to human health and safety thus appeared to be virtually nonexistent. He went on to say that no non-thermal or other low-level effects of ELF fields had been proved, and he dismissed the thirty-two studies of ELF effects cited by Becker and Marino as either irrelevant, inconclusive, or incompetently conducted.

Another witness who testified for the utility companies at the PSC hearings was Professor Sol M. Michaelson, a doctor of veterinary medicine at the Department of Radiation Biology and Biophysics of the University of Rochester School of Medicine and Dentistry. During the 1960s, Michaelson, whose research was largely financed by the Navy, had conducted numerous experiments on the heating effects of microwaves by exposing dogs and other test animals to radiation of sufficiently high intensity to produce deep burns and death within a few hours. Although he had performed few, if any, experiments at levels below ten milliwatts per square centimeter, he claimed to have confirmed the safety of Schwan's proposed standard.

In 1973, Michaelson submitted testimony to the Senate Commerce Committee, supporting a proposal by the Navy to study the effects of Sanguine ELF fields in human volunteers. "It appears entirely possible that these fields are involved in the etiology of certain human illnesses which have increased spectacularly during the last century," he told the committee. However, when he testified in behalf of Niagara Mohawk and Rochester Gas & Electric at the PSC hearings, he declared that ELF fields from power lines were safe, and warned that Soviet investigations showing ELF effects should be "viewed with caution." (Soviet scientists had determined that workers in high-voltage substations were experiencing altered pulse rates and blood pressure, as well as fatigue, drowsiness, and headache. As a result, the Soviet government had ruled that 750,000-volt transmission lines must be at least 100 meters from inhabited dwellings—200 feet more than the safety zone that had been proposed for the Massena-to-Marcy line.) As for the thirty-two published studies demonstrating ELF effects, Michaelson suggested that it was "possible to postulate a large number and variety of factors other than exposure to the electrical field as the cause of the observed difference." He also declared that a biological effect did not necessarily constitute a biological hazard, thus implying that an effect could be ignored if it had not been proved to be a menace to health.

A third witness for the utility companies was Professor Morton W. Miller, a botanist, who, like Michaelson, was a member of the Department of Radiation Biology and Biophysics at the University of Rochester School of Medicine and Dentistry. Miller, who had been a student of Michaelson, had conducted a three-year study for the Navy's research program on the biological effects of ELF radiation from Sanguine, in which he had determined that bean plant roots were not affected by exposure to ELF. In the testimony he filed for the PSC hearings, he stated that the proposed 765,000-volt transmission lines "did not pose an unreasonable risk to human health," basing this conclusion largely upon his analysis of studies of Sanguine that, like his own, had produced negative results.

Cross-examination of witnesses began in April 1976 in a state office building in Syracuse. At the outset, Robert Simpson, the lawyer for the Public Service Commission, forced Miller to concede that since the electric field produced by a 765,000-volt line was far stronger than the field emanating from Sanguine, negative studies of ELF radiation from Sanguine could not be used to demonstrate that the field from the power line would be safe. He then got Michaelson, who succeeded Miller at the witness stand, to concede that his position regarding ELF radiation from the proposed power line was not "what we don't know can hurt us," or "what we don't know can't hurt us," but "what we don't know *might* hurt us"—a rather questionable position to take in a hearing concerning the public health.

Michaelson was followed to the stand by Herman Schwan, who had testified in 1975 as an expert for the General Electric Company in hearings conducted by the Food and Drug Administration's Bureau of Radiological Health after the bureau had ordered the recall of 36,000 GE microwave ovens suspected of leaking unacceptable levels of microwave radiation. The bureau's final report of the case contained an appendix entitled "Undocumented and Unsubstantiated Statements Made by Schwan," which listed more than a hundred such statements. During cross-examination, Simpson forced Schwan to admit that he did not know how or whether the low-intensity current induced in bone by power-line radiation might affect bone growth, the healing of fractures, or the production of bone tumors. Simpson then got Schwan to acknowledge that he could not say how or whether ELF fields from transmission lines might affect other organs of the body, and Simpson asked him how, if he couldn't answer these questions, he could be certain that the proposed line was

safe. Finally, toward the end of the second day of cross-examination, Schwan said that he did not wish to continue answering
such questions. At that point, Simpson told the administrative
judge, who was presiding over the hearing, that he had no more
questions to ask.

During late April, Marino was cross-examined for eight days,
and early in May, Becker underwent cross-examination for four
days. Although attorneys for Rochester Gas & Electric and
PASNY claimed that Marino was prejudiced against power companies, and that Becker's testimony should be thrown out because it contained "the rankest form of hearsay," neither man
could be shaken from his conviction that because ELF fields had
been shown to produce biological effects in animals, they posed
a potential health risk for humans. By now, however, Becker and
Marino were not only locked in a no-holds-barred struggle with
PASNY and the utilities, but were also engaged in a bitter debate
with the National Academy of Sciences. This came about when
they learned that the Academy had appointed a committee of
experts to look into the biological effects of ELF radiation, and
that its members included Schwan, Michaelson, and Miller.

# SOME CONFLICTS
# OF INTEREST

DURING 1973 AND 1974, determined political opposition had driven Project Sanguine from Wisconsin, as well as from a backup site in the Texas hill country. In 1975, in an effort to make its ELF communications system more palatable to the public, the Navy proposed to reduce the size of the area the system would encompass from 22,500 square miles to between 3,000 and 4,000 square miles, and renamed it Project Seafarer. The Navy then suggested moving Seafarer to the Upper Peninsula of Michigan, where, because of high unemployment and a depressed economy, it was thought that the project might stand a better chance of being accepted. Tentative approval was given to Seafarer by Governor William G. Milliken of Michigan, but later he demanded the right to veto installation of the system after he learned that Navy officials had failed to inform his science advisers about the Pensacola study showing that exposure to ELF magnetic fields had raised triglyceride levels in humans, or about the other biological effects of ELF radiation which had alarmed Becker and the members of the Navy's 1973 advisory committee.

The report of the 1973 committee did not come to light until December 1975, when a copy of it was acquired by Senator Gaylord Nelson of Wisconsin. Nelson was furious at learning that the Navy had withheld the report for two years while some of his constituents were undergoing exposure to ELF at the test facility at Clam Lake. In a press release, he accused the Navy of suppressing evidence of a potential health risk to the citizens of his state, and of failing to perform the follow-up studies of ELF that had been urgently recommended by Becker and the other members of the advisory committee. Seeking to mollify Nelson and to defuse a politically embarrassing situation, the Navy offered the National Academy of Sciences a contract to conduct an inquiry into the possibility that plants, animals, and people might be harmed by the electric and magnetic fields associated with Project Seafarer.

Early in January 1976, the Academy announced that it had appointed sixteen experts to a Committee on Biosphere Effects of Extremely Low Frequency Radiation, which would be headed by J. Woodland Hastings, a professor in the Department of Biology of Harvard University. The list of candidates from which committee members were chosen had been drawn up by members of the National Academy's professional staff, who were said to have relied upon advice from the Navy in compiling it. The list was then presented to Hastings, an affable man with no previous experience in ELF research, whose selections for membership were subject to the approval of high Academy officials. Among them was Academy president Philip Handler, a biochemist known for his work on amino acids. In keeping with Academy policy at the time, only after the committee had been appointed were its members asked to fill out bias statements indicating what views they might have expressed publicly on the biological effects of ELF, or whether they might have other conflicts of interest.

When Marino learned that Schwan, Michaelson, and Miller had been named to the Academy committee, he telephoned Hastings to protest that the three men had already submitted testimony to the PSC hearings in New York that ELF radiation from a proposed power line would not harm the health of humans or pose a hazard to the environment. According to the account written by Marino and Joel Ray in "The Electric Wilderness," Hastings expressed surprise at this, and promised to have him and Becker appointed to the committee in order to give it better balance. Some weeks later, Hastings wrote Marino that he was "desperately trying to get approval" for Marino's appointment. Meanwhile, the Academy invited Becker and Marino to attend as guests and observers a public meeting of the committee scheduled for March 25. Instead of appearing at the meeting, Becker and Marino sent a memorandum to Hastings, quoting testimony given by Schwan, Michaelson, and Miller at the power-line hearings, and pointing out that because the electric fields from the proposed lines were far stronger than the fields from Seafarer, "it is inconceivable that the three named individuals will find that the [Sanguine/Seafarer system] is an environmental hazard, regardless of the evidence adduced." Their argument was supported by a group of researchers from Michigan Technological University, in Kalamazoo, who urged the Academy to purge the committee of "all members who have gone on public record as believing that ELF radiation is not harmful."

At the time, Academy officials defended the makeup of the committee by pointing out that one of its members, Dr. W. Ross Adey, director of the Brain Research Institute of the University of California at Los Angeles, had opposed Schwan's views concerning the biological effects of ELF fields, and that two other members had served on a Sanguine evaluation panel in Wisconsin that had been critical of some of the Navy's studies of ELF. Adey was, in fact, the only person on the committee who had conducted research on the biological effects of low-level ELF fields. Most of its members had no previous demonstrated experience with ELF studies, and were thus at a disadvantage in assessing or challenging the opinions of Schwan, Michaelson, and Miller, who were widely regarded as experts in the field. Moreover, several members—among them George M. Wilkening, director of environmental health at the Bell Telephone Laboratories in Murray Hill, New Jersey—were staunch advocates of the ten-milliwatt standard.

Meanwhile, the Navy went ahead with secret plans to subject fifty-two human volunteers to ELF at the Naval Aerospace Medical Research Laboratory in Pensacola, Florida. Early in January 1976, high officials of the Navy and the Department of Defense decided that the human experiments would begin in the spring, providing that no serious effects resulting from ELF exposure were observed in a monkey study then under way at the Pensacola laboratory. On March 10, the Navy's Committee for the Protection of Human Subjects voted 11–0 to approve the human experiments, after being assured by researchers at the Pensacola laboratory that "the only known potential risk is a rise in serum triglycerides," which "may or may not be dangerous," and that the results of the experiments, "whether negative or positive, would apply to a wide range of environmental questions regarding human exposure to fields associated with high-tension lines, novel transportation devices [these, presumably, were high-speed magnetic trains] and common household wiring."

The Navy told prospective volunteers for the experimental program that the electromagnetic fields to which they would be subjected would be "similar to those you experience while walking beneath electric transmission lines around your house, or while shaving, drying hair, mixing food with an electric mixer, and while working around a number of common household tools and kitchen items." In order to obtain their informed consent, the Navy explained that "preliminary research has shown [that] some blood constituents may change when exposed to similar fields, and some behaviors such as reaction time might be influenced." In the end, the Navy decided not to go ahead with ELF experiments on humans because male monkeys exposed to ELF radiation at the Pensacola laboratory gained weight faster than unexposed control animals, and because exposed female monkeys showed lower triglyceride levels than the exposed males.

In August of 1977, the National Academy of Sciences released the final report of the findings of its committee on the biological effects of ELF radiation from Seafarer. The report read in large part like a replay of the opinions and testimony that had been presented by Schwan, Michaelson, and Miller at the PSC hearings. Its authors dismissed Marino's experiments showing stunted growth in rats and mice that had been exposed to ELF radiation, and concluded that "there should be no concern for possible effects of Seafarer on fertility, growth, or development." They found deficiencies in the studies demonstrating that ELF magnetic fields caused elevated triglyceride levels in humans, and concluded that "Seafarer fields will not have an effect on human triglyceride concentrations." They passed quickly over findings that ELF radiation caused slowed responses in humans and monkeys, and, ignoring a 1974 report showing that ELF magnetic fields had degraded the ability of human volunteers at the Naval Aerospace Medical Research Laboratory to perform simple addition, stated their belief that the behavioral effects of ELF "do not appear to warrant concern."

In the end, they found themselves able to assure the public that, except for possible electric shocks, "the likelihood of serious adverse biologic effects of Seafarer is very small." Moreover, after claiming at the outset of their report to be "aware of the larger question" of the potential hazards that might be involved in the exposure of human beings to ELF fields, such as those emanating from power lines, the members of the Academy committee had not one conclusion to draw, nor a single recommendation to make, nor, indeed, another word to say about this question, except to point out in passing that "power lines sometimes carry currents much larger than those of Seafarer."

Chapter

# SOME WIDELY ACCEPTED RISKS

THE LONG-DRAWN-OUT PSC HEARINGS in New York had ended in March 1977, following a grueling cross-examination of Marino by attorneys for the utility companies and PASNY. They charged that Marino was biased and irresponsible because, unlike Michaelson and Miller, he had allowed himself to be interviewed about the power-line controversy by Mike Wallace of the CBS Show "60 Minutes," and because he and Becker had written a letter to Governor Carey in which they had described PASNY's 765,000-volt transmission line as an "imminent health hazard." Administrative Law Judges Thomas R. Matias and Harold L. Colbeth, who were presiding over the PSC hearings, were highly critical of Marino's conduct, which they considered to be inappropriate on the part of someone who was giving testimony at a public inquiry. As a result, when the National Academy's report on Seafarer was released four months later, the two judges reopened the hearing record and allowed the report to be introduced as evidence. According to the judges, this was done in order to show that "distinguished, nationally known" scientists on the committee had disagreed with Marino's opinions and conclusions regarding ELF.

Matias and Colbeth had already arranged for Dr. Asher R. Sheppard, a physicist at the Brain Research Institute at UCLA, to evaluate the conflicting testimony that had been given at the hearings, and to help them write the recommended decision they were now required to hand down. Sheppard himself had not performed any experiments with ELF fields. However, together with Professor Merrill Eisenbud, director of the Laboratory of Environmental Studies at the New York University Medical Center's Institute of Environmental Medicine, he had recently conducted a review of the literature on the biological effects of ELF fields for the American Power Service Corporation, and come up with a balanced appraisal of the problem. On the one hand, he and Eisenbud found "no evidence that the public health or ecological systems have been jeopardized in the slightest by artificial electromagnetic fields." On the other hand, "The one firm conclusion that emerges from a review of the existing literature is that relatively weak electric or magnetic fields are capable of evoking neurophysiological or behavioral effects" in monkeys and humans. They went on to suggest that because of "existing uncertainties and questions raised by the Soviet reports, it is important that a properly conducted epidemiological study be undertaken."

The call for further studies before preventive measures was typical of Professor Eisenbud's general approach to the problem. Eisenbud was a charter member of the Electromagnetic Radiation Management Advisory Council (ERMAC), which had been set up to advise the President about the biological effects of microwave and radio-frequency radiation. In October 1978, at a public hearing held by the New York City Board of Health, he would declare his opposition to a proposed environmental standard for exposure of the general population of the city to such radiation, saying, "We need several years of additional research, perhaps even a decade, before a new standard is set." Shortly thereafter, he told an ERMAC meeting that the proposed environmental standard was a "bad recommendation," and that the city Health Department "should have said there was no problem." At the same meeting, he left little doubt about whom he blamed for the increasing public concern about exposure to microwave and radio-frequency radiation. He even proposed a solution. "The scientific community should square off against the press," he declared. "Scientists should attack the press."

Soon after the National Academy released its 1977 report exonerating ELF radiation from Seafarer, Marino filed an *amicus curiae* brief in the power-line case in which he urged the PSC to take protective measures not only against ELF fields emanating from the proposed 765,000-volt power lines, but also against fields that were being given off by existing 345,000-volt lines. In his brief, Marino maintained that "the scientific literature which shows that the proposed transmission lines would be a human health hazard also shows that existing high-voltage transmission lines *are* a human health hazard." He based his argument on the fact that he and other researchers had found biological effects in test animals exposed to the equivalent of 1,500 volts per meter, whereas the electric-field strength at the edge of the 150-foot right-of-way of a standard 345,000-volt line had been calculated to be approximately 1,600 volts per meter. Marino proposed that a safety factor of at least one hundred be applied, and that the upper limit of permissible chronic human exposure be set at 100 volts per meter. He said that to permit exposure to more intense fields would be tantamount to subjecting people to involuntary human experimentation, and he suggested that doing so could make power companies liable for the tort of battery, or for the taking of constructive easement over land. The average transmission line carries only a portion of its deliverable electric power within its right-of-way, Marino pointed out, and as much as 50 percent of its power exists in measurable field strengths beyond the right-of-way. Marino warned that people living in the vicinity of a 765,000-volt power line would be exposed to a strong alternating-current magnetic field that would be "totally new and unique with respect to the evolutionary history of life on earth." Finally, he recommended that the PSC require PASNY and the utilities to inform the people of New York State of the existence of a valid scientific dispute concerning the hazard posed to human health by exposure to electric and magnetic fields from high-voltage transmission lines, and he urged the PSC to organize a major research program conducted by independent scientists, in order to determine the extent of the hazard.

On January 20, 1978, Matias and Colbeth handed down their recommended decision—a 156-page document that reflected considerable ambivalence on their part. On the one hand, nearly a third of it was devoted to a blistering attack upon the credibility, competence, and conduct of Andrew Marino, who was accused of having been "evasive and argumentative under cross-examination," and whose experimental work was described as "not conducted carefully enough for the results to be believable." On the other hand, Matias and Colbeth conceded that ELF radiation might cause biological effects, and agreed with Becker and Marino that effects not known to be benign must be considered to be potentially hazardous. With respect to the proposed power line, the two judges concluded that "continuous long-term exposure to electric fields exceeding 2500 volts per meter might result in some biological effects that might be harmful." After selecting a safety factor of only two and a half, they went on to recommend that no person should live or work regularly in areas where the electric field exceeds 1,000 volts per meter at one meter above the ground. This meant that a 765,000-volt power line would require a right-of-way approximately 350 feet wide, instead of the 250-foot-wide corridor that had been proposed by PASNY and the utility companies.

Matias and Colbeth saw no need to take any action with respect to transmission lines carrying voltages of less than 765,000 volts. Nor did they see any need to post warning signs on power-line rights-of-way, or to direct the utility companies to finance biological research into the power-line hazard, as both Marino and the PSC staff had urged. They did, however, recommend that the public not be encouraged to use the rights-of-way of 765,000-volt lines for recreational purposes—something the PSC had allowed for many years in the corridors beneath 345,000-volt lines, which furnish trails for snowmobilers, cross-country skiers, and hikers, and have been leased by some communities to provide playgrounds for children and playing fields for Little League baseball teams.

As might be expected, the decision of the administrative law judges pleased none of the parties to the hearing. PASNY and the utility companies objected to the recommendation that would require them to create wider rights-of-way, and PSC staff members faulted them for omitting any detailed consideration of the thirty-two studies Marino had cited. The PSC staff claimed that nine of these studies provided "a solid body of evidence that electric fields from the transmission line will probably cause biological effects in humans," and for this reason, they recommended that the utilities be required to inform each resident living adjacent to the right-of-way of a 765,000-volt line that there was a risk from exposure to the overhead lines. They also recommended that anyone living within 275 feet of the centerline of the right-of-way of a 765,000-volt power line be given an option to have his house purchased or relocated if the electric field outside the dwelling exceeded the 400-volt-per-meter level.

The final decision in the power-line case came in June 1978, when, after reviewing the hearing record and the judges' recommended decision, the six members of the Public Service Commission issued a seventy-four-page opinion. It straddled the public health issue. To begin with, the commissioners found that although the hearing record did not show that the electric and magnetic fields of the proposed 765,000-volt lines would endanger human health and safety, "it contains unrefuted inferences of possible risk that we cannot possibly ignore." They based their assessment of possible risk on Marino's three-generation mouse study, and on the other studies showing that ELF fields produced biological effects in test animals, and they declared that "these effects cannot be presumed harmless." They went on to dismiss allegations that Marino's experiments were flawed, pointing out that PASNY and the utilities "could have aided the record by attempting to replicate Dr. Marino's experiments in a manner free of the defects they perceived in them," but that "for reasons best known to them, they did not do so." As for the harsh criticism leveled against Marino by Matias and Colbeth, the commissioners let stand the finding of their staff that PASNY and the utility companies had mounted "vigorous efforts to discredit him [that] were more suitable to a slander trial than a fact-finding scientific investigation."

It was one thing for the commissioners to agree with Marino and Becker that ELF fields could cause biological effects in test animals, however, and quite another for them to decide how far to go in protecting the public health against the potential threat posed by power-line radiation. Unlike Matias and Colbeth, they could not very well ignore the question of how safe the existing 345,000-volt transmission lines might be, when their own staff had recommended a 400-volt-per-meter level for maximum human exposure, and when people living in houses adjacent to the rights-of-way of 345,000-volt transmission lines were already being exposed to electric-field strengths well above that level. The staff proposal posed a quandary for the commissioners. They were already under pressure from the utilities to make certain that power lines operating at lower voltages would be exempt from any standard they might adopt for 765,000-volt lines, and they apparently feared that if they adopted the 400-volt level, or the staff's recommendation that people living adjacent to the power-line right-of-way be warned of the possibility of a health risk, they might create widespread panic and a public furor. In the end, they attempted to solve the problem by declaring that the right-of-way of a 765,000-volt transmission line should be wide enough so that the strength of the electric field at its edge was no greater than the electric-field strength (about 1,600 volts per meter) that had been calculated to exist at the edges of the rights-of-way of existing 345,000-volt lines. "In this way," they reasoned, "we assume that the risks, if any, of long-term exposure to 765 kV transmission in the areas traversed by PASNY's line and any future 765 kV lines will be no greater than those, now widely accepted, of long-term exposure to the 345 kV lines operating throughout the State."

The sweeping supposition of the phrase "now widely accepted" appeared to trouble the commissioners, for they acknowledged in a footnote that "we do not imply that society has, in any way, explicitly decided that 345 kV lines are worth whatever risks they may entail." Whatever the commissioners wished to imply, the fact remained that both the assumption and the qualifying footnote were unfounded. Thanks to inadequate coverage of the power-line hearings by the media, the public had been given almost no information about the risk posed by power-line radiation, let alone enough to reach any kind of rational decision. Anyone who chose to read to the end of the PSC's opinion, however, would have come upon a dissent filed by Commissioner Harold A. Jerry, Jr., who put an astute finger on the motive behind the decision of his five colleagues to extend the presumption of benignity to 345,000-volt power lines.

"I believe that the majority has settled for this width because of irrelevant concern over the widths of existing 345 kV rights-of-way, rather than because of concern for the proper widths of 765 kV lines in view of the evidence in this case," Jerry wrote. "In other words, the majority has been swayed by the problem of what to do with existing 345 kV lines if it adopts a 765 right-of-way with weaker field strengths at its edges than those present at the edges of existing 345 kV rights-of-way."

In spite of temporizing on the question of the safety of 345,000-

volt lines, the PSC opinion was a landmark decision, marking the first time that any regulating body in the nation had gone on record as stating that studies showing that ELF fields could cause biological effects in animals must raise questions about human exposure to such fields along power-line rights-of-way. Moreover, the commissioners were sufficiently troubled about the many unanswered questions about ELF radiation to take the unprecedented step of calling for a five-year research program on the biological effects of power-line radiation, to be financed by PASNY and the utility companies and carried out by independent scientists.

A few months after the commissioners issued their decision, the Power Authority brought suit against the PSC, claiming that the Commission had no right to require it to participate in such a program as a condition of certifying a transmission line. In 1979, the appellate division of the New York State Supreme Court upheld the Power Authority's position in the matter. But in 1980, while that decision was on appeal to the State Court of Appeals, the PSC and PASNY reached a settlement under which the Power Authority and seven major electric utility companies agreed to provide $5 million to finance a five-year research program that would be known as the New York State Power Lines Project. A three-member board made up of the commissioner of the New York State Health Department, the chairman of PASNY, and the chairman of the PSC was given responsibility for establishing the project, and early in 1981 the board selected a ten-member scientific advisory panel to develop the project's research program and oversee its research contracts.

# RETALIATION

ON THE FACE OF IT, the decision of the Public Service Commissioners and the subsequent settlement of the legal dispute between PASNY and the PSC seemed to indicate that Becker and Marino had finally prevailed in their long struggle to gain official acknowledgment of the biological effects and potential health hazard of 60-hertz radiation from power lines. As a result of their stand, the Power Authority would soon announce that it intended to postpone indefinitely the construction of any more 765,000-volt lines, and Rochester Gas & Electric and Niagara Mohawk would eventually be denied permission to build their proposed 765,000-volt line from Rochester to Oswego.

In the meantime, Becker, Marino, and Maria Reichmanis, a graduate student in Becker's laboratory, had participated in a pioneering study of suicide with Dr. F. Stephen Perry, an English physician from the West Midlands. Perry had initiated the investigation after observing during the course of his general medical practice that people living near high-voltage transmission lines appeared to suffer from a higher than normal incidence of depressive mental illness. In 1979, using computer-based calculations, the four investigators reported in *Physiological Chemistry &*

*Physics* that they had found a correlation between the presence of electromagnetic fields from power lines and the occurrence of suicide. Two years later, after measuring actual magnetic-field strengths at the homes of 590 suicide victims and of 594 people who served as a control population, they reported in *Health Physics* that "Significantly more suicides occurred at locations of high magnetic field strength." They believed this to be "the first demonstrated correlation between human behaviour and environmental power-frequency fields," and they called for a large-scale epidemiological study to determine the public health significance of their startling finding. By this time, however, Becker and Marino had become the target of savage criticism and had been forced out of the Veterans Administration.

In 1964, Becker had won the Administration's coveted William S. Middleton Award for Outstanding Achievement in Medical Research, and in 1971 the VA had granted him the status of medical investigator, which meant that he could put aside his clinical duties and devote his time and energy to research. In April 1976, however, just as he was about to take the stand for the first time in the power-line hearings, he was informed by the VA's assistant chief medical director for research and development in Washington, D.C., that his application for a five-year renewal of his medical investigatorship had been deferred. Later that year, a research grant he had been receiving from the National Institutes of Health was discontinued, and from then on he began to experience great difficulty in getting his research projects approved by the VA.

Becker managed to keep his laboratory going for another three years by finding an alternative source of financing at the regional level of the VA. In 1980, a year after he won the Nicolas Andry Award of the Association of Bone and Joint Surgeons for outstanding achievement in the field of orthopedic surgery, all his VA grants were terminated, and Becker retired from the Administration and went to live in Lowville, where he presently directs a research and consulting firm called Becker Biomagnetics. Six months later, his research laboratory, which had been operating for twenty years, was shut down. Shortly thereafter, Marino, who had worked in the lab for seventeen years, was told that there were no suitable positions for him at the hospital. As a result, he resigned from the Veterans Administration and went to Louisiana, where he is now a professor in the Department of

Orthopedic Surgery at the Louisiana State University Medical Center in Shreveport.

Although the motives behind the apparent vendetta against Becker and Marino can only be guessed at, it seems clear that their outspokenness regarding the health hazard of power-line radiation must have had something to do with it. Moreover, after the 1978 PSC decision, the two men came under attack from other quarters. In 1979, Professor Hastings, who had chaired the National Academy of Sciences' committee on Seafarer, told Susan Schiefelbein, who was writing a piece about the ELF controversy for *The Saturday Review,* that Marino and Becker had not conducted any significant research on the biological effects of ELF, and that the judges in the power-line hearing had decided against them. "The judges threw out the case with prejudice," Schiefelbein quoted Hastings as saying. "[They] ruled that Marino's not a believable witness, that he's evasive and deceitful. Here we were, being attacked by people who ultimately were thrown out of a court of justice in that way. They've all been thrown out. These guys are all a bunch of quacks."

Another attack on Marino was mounted by National Academy president Philip Handler, after Schiefelbein's article had appeared in *The Saturday Review* in September 1979. Her article was entitled "The Invisible Threat: The Stifled Story of Electric Waves," and it was highly critical of the makeup and conduct of the National Academy's committee on Seafarer, and of the credibility of Professor Hastings. A few months after publication, Handler wrote to Carll Tucker, the editor of *The Saturday Review,* threatening to sue the magazine unless Tucker published a sixteen-page article entitled "Scientific Evidence and Public Decision Making" which Handler had written with the assistance of two senior officials of the National Academy.

In the article, Handler and his colleagues criticized Schiefelbein for "failing to note that Dr. Schwan, a member of the National Academy of Sciences, is perhaps the leading authority in the United States, if not the world, on the interactions of electromagnetic fields within living tissue." They then charged that she had deliberately disregarded the fact that Marino's experiments had been "rejected as valueless by the rules by which science guards against shoddy work." It soon became apparent that the rules they were referring to had been laid down by Schwan, Michaelson, Miller, and the other members of the National Academy's committee on Seafarer. "The Committee's reviewers found that the cages used to house [Marino's] experimental animals could have transmitted small electric shocks each time the rats ate or drank," Handler and his co-authors wrote. "Was it then these shocks or the [electromagnetic] fields that led to poor feeding by some rats? Did Marino consider such shocks in his conclusions? One doesn't know but it seems likely that to be 'buzzed' when one eats is not to eat well."

If Handler and his colleagues had given the Academy's own report on Seafarer a careful reading, they would have learned that Marino and his co-workers had been the first to point out that microshocks might have influenced the results of their experiment. Moreover, if they had read the original report of the study, they could have learned that Marino and his co-workers had investigated and discounted the possibility that microshocks had been a factor. Even more important, if they had read page 24 of the Public Service Commission decision, they would have come across the following paragraph:

> Applicants' witness Miller made movies of rats in cages constructed to replicate Marino's apparatus and claimed that he could observe the rats recoiling when they drank while the [electric] field was on; staff reviewed the movies and contends that they "demonstrate unequivocally that there is no difference in the behavior of rats drinking in either the 'field off' or 'field on' condition."

After acknowledging that "an electric field is created within a person standing under an electrical transmission line," Handler and his associates pointed out, as Schwan had done at the powerline hearings, that this internal field was "thousands or more times smaller than the external field in the air." They declared that while "plentiful data" had been collected to determine whether weak internal fields could pose a health hazard, "much of them are contradictory, and some simply experimentally invalid." They followed this sweeping and largely undocumented assertion by concluding that "if a hazard does exist it has not been demonstrated," and that in the absence of proof of a hazard, or of any acceptable theory predicting a hazard, "there does not exist any danger from extremely low frequency radiation at the level at which people are customarily exposed.'' They then lashed out at the press for using "the thin tissue of fancied biological effects of ELF to inflame the imagination of the public.''

In the end, Tucker stood by the article that Schiefelbein had written, and the piece by Handler and his associates went unpublished in *Saturday Review*. Handler's article, however, exemplified the thinking with regard to the power-line problem at the highest level of the nation's scientific establishment at the time that Nancy Wertheimer discovered the association between childhood cancer and exposure to alternating magnetic fields from ordinary electrical distribution lines—a finding she hoped that the medical and scientific community would be eager to investigate in greater detail.

# OBFUSCATION

In October 1979, Wertheimer, who was about to become a grand-mother for the first time, drove east to be with her daughter when the baby was born. She had never met Robert Becker or Andrew Marino, but she had recently cited their studies of ELF radiation effects in her application for a National Institutes of Health grant to study the possible association between cancer in adults and exposure to magnetic fields from high-current wiring, so she stopped over in Syracuse and paid them a visit in their laboratory at the VA Hospital. While there, she learned that the National Advisory Environmental Health Sciences Council—a senior group that was conducting a final review of her proposal—had sided with the minority on an earlier review panel and turned down her application. "I was disappointed, of course, but I didn't let it get me down," she said recently. "As it happened, I had been gathering data for the adult study for almost a year, so I just kept right on going and did it the way I had done the childhood study—on my own time and with my own money."

To begin with, Wertheimer collected death certificates for people aged nineteen to sixty-two who had died of cancer between 1967 and 1975 in the towns of Boulder and nearby Longmont, and also 1977 cancer death certificates for residents of the city of Denver and its suburbs. These four samples also included cancer survivors—people who had been diagnosed five or more years earlier as having a life-threatening form of cancer, and who were still living in 1979 without known recurrence of the disease. In each of the samples, only those cancer victims who had lived in the sampling area for at least four years prior to diagnosis were retained for the study. Control addresses for the cancer victims from Boulder and Longmont were selected from non-cancer death certificates in the same towns, and the control addresses for the Denver City and suburban Denver cancer cases were chosen at random from within the same neighborhoods. By the spring of 1979, Wertheimer had collected addresses for 1,179 cancer victims and the same number of controls, and for the rest of that year and most of the next, she drove around—often from sunrise to sunset—visiting case and control homes and classifying each of them according to her wiring configuration code, which she had refined somewhat since the childhood study.

"One reason I had hoped to get an NIH grant was so that I could afford to hire someone to do the coding of the addresses blind—in other words, someone who, unlike myself, would be totally unaware of which addresses had been the residences of cancer victims and which were the homes of controls," she says. "In that way, I could have insured that no bias could creep into the final results because of non-blind coding. Without financial support, however, that was impossible, so I hired a researcher out of my own pocket to do blind coding of one hundred and forty of the eleven hundred and seventy-nine case-control pairs. When it turned out that the researcher's blind coding did not differ significantly from my own coding, and that a significant result was obtained from the blind-coded sample alone, I was reassured about using my not-blind coding in the final analysis of the data."

While Wertheimer was busy gathering information for her adult study, the electric utilities industry was trying to decide how to respond to the disturbing findings of her childhood study. In July 1979, she and Leeper were visited in Boulder by Robert Kavet, the project manager of biomedical studies for the Electric Power Research Institute (EPRI), an organization situated in Palo Alto, California, that is sponsored by major utility companies across the nation. During his stay, Kavet discussed the childhood cancer study at considerable length with Wertheimer and Leeper, and accompanied them to several neighborhoods in Boulder, where they used a gaussmeter to measure magnetic-field strengths outside a number of homes

When Kavet returned to Palo Alto, he wrote a report about his visit, and sent it on July 30 to EPRI's member companies. In the report, he noted that Wertheimer and Leeper had not measured magnetic fields inside the homes they had studied. Kavet also described the dilemma that their findings posed for the electrical utilities industry. On the one hand, he wondered how magnetic fields in the milligauss range could possibly correlate with increased cancer rates in the absence of scientific evidence to support the existence of such a relationship. On the other hand, he pointed out that the childhood cancer study had been peer-reviewed and published in the highly esteemed _American Journal of Epidemiology_. He added that he believed that Wertheimer and Leeper "have attempted to be objective with their data and are _not_ out to achieve a media 'splash.' "

As for how EPRI's member companies should respond to the findings, Kavet offered three possible approaches. His first suggestion was to take readings of indoor magnetic fields in a representative number of homes in the childhood cancer study to determine whether strong magnetic fields were being given off by household appliances—a possibility that the Navy had suggested in its 1972 environmental impact statement for Project Sanguine. His second suggestion was to take a careful look at the small body of medical and scientific literature suggesting that electromagnetic interaction with biological systems could occur at extremely low field strengths. His third suggestion was: "We can do nothing." Kavet ended his letter by asking for reactions, saying, "I would appreciate your feedback. This is a complex issue, and one that has stirred interest and controversy. I do not believe that any single person has the solution, but collectively we might be able to pool our insights."

On August 10, Wertheimer wrote to Kavet and thanked him for sending her a copy of the report. "As you say, the question of what to do next is complex," she agreed. "I rather doubt that much will happen so long as our one study is the only suggestive evidence in a sea of disbelief. However, we'd be glad to help in any way we could if you decide on a study project." At the end of her letter, Wertheimer cautioned Kavet regarding his suggestion that the electric utilities industry measure in-house magnetic fields. "You recall the misunderstanding (which you shared with us and many others) about the magnetic fields from appliances as reported in the Sanguine studies," she wrote. "Their magnetic-field readings were taken as close as possible to the appliance housings, and so give a highly exaggerated idea of what ambient fields might be expected within the homes from appliance sources. I'm enclosing our correspondence with them on this matter."

The enclosed correspondence consisted of a letter that Ed Leeper had written in February 1978 to Donald A. Miller of the Naval Underwater Systems Center in New London, Connecticut, with a copy to Anthony R. Valentino, manager of electromagnetic effects at the Illinois Institute of Technology Research Institute (IITRI) in Chicago, Illinois, and a reply that Valentino had written to Leeper in behalf of himself and Miller. (Miller had worked at IITRI when it was under contract to the Navy to produce the 1972 final environmental impact statement for Project Sanguine and, together with Valentino, he had helped write the statement before going to the Naval Underwater Systems Center.) In his letter, Leeper reminded Miller that although the impact statement declared that IITRI had measured very strong magnetic fields in the vicinity of ordinary consumer appliances, such as hair dryers, television sets, and fluorescent lamps, Miller had subsequently acknowledged in a telephone conversation with Leeper that he had been able to measure strong magnetic fields only when he held the detector coil close to each appliance.

Early in March 1978, Valentino wrote to Leeper. "I have discussed your letter with Dr. Miller," he noted, "and we both agree with your concern over the possible misinterpretation of measurements taken as near as possible to various appliances, as compared with other measurements taken at some distance." In 1983, however, almost two years after President Reagan and Secretary of Defense Caspar W. Weinberger had ordered the Navy to proceed with the development of Project ELF (as Sanguine/Seafarer was now called), IITRI and the Navy issued a report entitled "Representative Field Intensities near the Clam Lake ELF Facility," which claimed once again that strong magnetic fields up to 25 gauss had been measured at "user distances" from hair dryers and other household appliances. In a cover letter to the report, which was sent to interested members of Congress, Rear Admiral Bruce Newell, chief of legislative affairs for the

Navy, assured the lawmakers that all of the electromagnetic fields that had been measured as coming from the Clam Lake ELF antennas "are well below that level which would be of concern to humans or the environment."

A few days later, an assistant to Senator Carl Levin of Michigan sent a copy of the IITRI report and the admiral's cover letter to Wertheimer and Leeper, and asked for their comments. Wertheimer and Leeper expressed their reaction in a letter dated April 11, in which they enclosed Leeper's 1978 correspondence with Miller and Valentino. "We are concerned that this report by IITRI and the Navy, and the Navy cover letter that accompanies it, are misleading on several points," they wrote. "We are particularly concerned because we know that some of the errors contained in the report are repetitions of errors reported elsewhere that have been previously (several years ago) called to the attention of IITRI."

Wertheimer and Leeper went on to point out that the Institute's measurements of magnetic fields produced by household appliances "were not taken at 'user distances' as most people would understand that phrase," but were made "by using a very small probe to search out the maximum fields right up against the casing of the appliances." This resulted in readings that were "much higher than those which could be considered the real exposure to the user of the appliance." Wertheimer and Leeper then characterized as "simply not true" Admiral Newell's assertion that radiation from Clam Lake ELF antennas was "well below that which would be of concern to humans or the environment," pointing out that the magnetic fields they had found in homes with an increased cancer risk had averaged about one milligauss. By contrast, a level almost two and a half times that high had been measured at a distance of one mile from the ELF antennas at Clam Lake. The two researchers concluded their letter by observing that the Navy appeared to be trying to reassure the public that ELF radiation was harmless, but that "presenting information in a way which would tend to mislead those with the responsibility for making decisions is not, in our opinion, the best way to reassure the public."

Wertheimer and Leeper sent copies of their letter to Senator Levin's assistant; to Senator Faust of Michigan; to Governor Art Blanchard of Michigan; and to Senators Spike Owen and Don Riegle of Wisconsin—all of whom reportedly expressed strong disapproval to the Navy about the erroneous information in its report on the Clam Lake ELF facility. Not until 1984, however, did IITRI prepare a report for the Naval Electronics Systems Command that finally showed (as Wertheimer and Leeper had been claiming for nearly a decade) that magnetic fields from household appliances dropped off so rapidly that they could not be considered as serious sources of indoor magnetic-field exposure. By that time, the Navy had received permission to rebuild and reactivate the ELF transmitter at Clam Lake, and to construct an ELF transmitter near the K. I. Sawyer Air Force Base in the Upper Peninsula of Michigan.

Chapter

# IN A BIND

IN SPITE OF THE EFFORTS of Wertheimer and Leeper to set the record straight, scientists supported by the utilities industry tried on several occasions during the early 1980s to discredit their theory about the relationship between high-current wiring and cancer by resurrecting the premise that strong magnetic fields were given off by ordinary household appliances. One of the first to do so was Professor Morton W. Miller, of the University of Rochester, who had testified at the New York power-line hearings as a paid consultant of Rochester Gas & Electric and was a member of the National Academy's committee on Seafarer. In July 1980, the *American Journal of Epidemiology* published a letter from Miller pointing out that Wertheimer and Leeper had failed to provide data on magnetic-field strength inside the homes they had surveyed. Miller went on to suggest that the magnetic fields inside these homes did not come from nearby electrical distribution wires, but "were dominated by ambient household fields." He then cited data presented by Donald A. Miller to argue that household appliances produced strong magnetic fields in household wiring. Professor Miller concluded by declaring that Wertheimer and Leeper's "suggestion of a causal relationship between cancer and the transmission line magnetic fields appears untenable.''

In a letter of reply published in the same issue of the *Journal*, Wertheimer and Leeper noted once again that ordinary household wiring does not have any significant effect on ambient magnetic fields within a house, because it consists of two parallel wires carrying equal currents in opposite directions—a configuration in which the magnetic fields generated by the currents "effectively cancel each other out." They added that Miller had overlooked the fact that significant magnetic fields can be generated in homes by unbalanced currents from nearby distribution lines, which often flow in home plumbing. They also noted that by citing data that misinterpreted the distance at which Donald A. Miller of IITRI had measured magnetic fields given off by household appliances, Professor Miller had greatly exaggerated the strength of these fields.

Four months after Professor Miller's letter was published, a second attempt to discredit the work of Wertheimer and Leeper was made by Kanu Shah, of Shah & Associates, a Gaithersburg, Maryland, engineering and consulting firm. In 1974, Shah had prepared a report for Rochester Gas & Electric stating that electromagnetic fields from power lines were not a health problem, and the utility had submitted it as evidence in the early stages of the New York power-line hearings. In November 1980, Shah wrote a critique of Wertheimer and Leeper's childhood cancer study that was presented at a board meeting of the Sacramento Utility District, whose members were considering where to place a 230,000-volt transmission line from Rancho Seco, a nuclear-powered electrical generating plant, in Sacramento County. After describing Wertheimer and Leeper's study as "non-scientific," Shah said that its findings had been contradicted by the authors of a study entitled "Electrical Wiring Configurations and Childhood Leukemia in Rhode Island," published that same month in the *American Journal of Epidemiology*.

Shah was correct that such a study had been conducted and published by an epidemiologist named John P. Fulton and some colleagues in the Section of Community Health at Brown University, in Providence. What Shah neglected to mention, however, was that the next issue of the *Journal* contained a critique of Fulton's study written by Wertheimer and Leeper. Fulton and his colleagues at Brown had been invited to respond by the editors of the *Journal*, but had declined to do so. In their critique, Wertheimer and Leeper pointed out that Fulton and his associates had collected birth addresses occupied during the 1950s as controls for addresses occupied by teen-aged cancer victims in the 1970s. As a result, the controls were biased toward urban addresses and, therefore, toward exposure to high-current wiring, which is known to be more prevalent at city addresses in Rhode Island than at suburban addresses. By doing this, Fulton and his colleagues had ignored the common tendency for young families to live first in rented city apartments or houses, where high-current wiring is prevalent, and then to move to the suburbs as their children grow up. When Wertheimer and Leeper reworked Fulton's data to reduce its urban bias, they found that leukemia had, indeed, occurred somewhat more often among Rhode Island children living in homes near high-current wiring than among children living in homes near low-current wiring. They then explained why the association between magnetic fields and childhood cancer was not as strong in Rhode Island as it had been in Denver:

> A large problem in researching environmental magnetic fields is obviously the ubiquitous exposure from sources other than the one studied (in this case, proximity of certain outdoor wires to dwellings). In Rhode Island, the distribution wires almost universally run along the streets, creating a relatively high exposure for anyone using the street or sidewalk. Such frequent street exposure may well "swamp out" effects due to exposure at the dwelling site. In Denver we were fortunate in that the wires were usually run along backyard lines, quite far from the street, so such background exposure is minimized.

In spite of Wertheimer and Leeper's unrebutted critique of the Fulton study, utility companies continued to tout the Rhode Island investigation as having refuted the theory that alternating magnetic fields could cause cancer in children, just as they continued to ignore the fact that household appliances were not capable of generating ambient magnetic fields of any significant strength. In the spring of 1981, Wertheimer addressed the motive behind this subterfuge in a letter she wrote to a man who had sent her a copy of Shah's report to the Sacramento Utility District. She pointed out that since the magnetic fields given off by ordinary high-current distribution lines were as strong as the magnetic fields existing at the edges of the rights-of-way of ultra-high-voltage transmission lines, "the power companies are really in a bind if they agree that magnetic fields might be harmful."

Chapter

# THE ELECTRIC POWER
# RESEARCH INSTITUTE

BY THIS TIME, the power companies had apparently decided that the best way out of their bind was to try to discredit Wertheimer. Early in January, EPRI awarded a contract to H. Daniel Roth of Roth Associates, Inc.—a consulting firm in Rockland, Maryland, that had done work for Consolidated Edison of New York and other utility companies—to reanalyze Wertheimer and Leeper's childhood cancer study. Shortly thereafter, Roth called Wertheimer and asked her if she would be willing to supply him with the data she had used in conducting the study. On January 7, she wrote him back (with a copy to Robert Kavet of EPRI) to say that she felt "uneasy" about the scope of his request, because she had limited resources and was very busy writing a report of her investigation of magnetic fields and cancer in adults. "I don't want to be uncooperative, and I believe that scientific data should be fully shared," she said. "But I honestly doubt that reanalysis of our data will yield any important new information. EPRI could use its funds more effectively, I should think, in financing a replication of our work, using whatever controls they deem appropriate." Wertheimer then told Roth: "I am left with the impression that, at present, EPRI's main concern is to find some procedural or statistical criticisms with which to refute our study, rather than to investigate vigorously the possibility of adverse magnetic field effects.''

An impression similar to Wertheimer's might have been formed by looking at the Institute's previous actions regarding the possibility that electromagnetic fields from power lines could pose a threat to human health. During the late 1960s, Soviet scientists had reported that electrical switchyard workers in the USSR were complaining of reduced sexual potency, and that the electric fields to which they had been exposed had an adverse effect upon the central nervous system and the heart. At about the same time, it was learned that Swiss engineers who were exposed to the low-frequency fields of electric railways were known to develop heart problems after the age of forty-five, and that because of this they were being retired at the age of forty.

In 1971, EPRI financed a five-year study of the biological effects of exposure to high-voltage electric fields. The study was conducted by Dr. Donald S. Gann, a neurologist at the Johns Hopkins University School of Medicine in Baltimore. Gann exposed anesthetized dogs to 60-hertz electric fields of 15,000 volts per meter for five-hour periods, after inducing small, carefully controlled hemorrhages in the animals in order to determine whether the fields would alter their response to the stress imposed by the hemorrhages. He found that the exposed animals experienced a significant decrease in blood pressure and a startling increase in heartbeat rate, as compared to unexposed control animals. In a 1975 report to EPRI, Gann described his findings as follows:

> The unexpected finding of these changes suggests strongly that dynamic effects resulting from exposure to electric fields may not be particularly subtle at all, but may be quite easy to detect. In addition to the findings with regard to magnitude of change, the variability of the heart rates of exposed subjects was also significantly greater than that in unexposed subjects, suggesting that the observations made by Soviet workers on conscious human beings exposed to high voltage electrical fields may be present in anesthetized dogs. These results are clearly preliminary but also clearly demand further explanation.

On June 1, 1976, Cyril L. Comar, director of EPRI's environmental assessment department, wrote Gann a letter informing him that EPRI was not going to renew his contract for the project. According to Comar, EPRI's Advisory Committee on the Effects of Electrical Fields—among its ten members were Professor Michaelson, two members of EPRI, and officials from five electric power companies—had decided unanimously that Gann's work did not warrant further investigation. Comar assured Gann that "this decision is in no way a reflection on your professional competence," explaining, "it appears that the heavy University demands on your time and abilities has made it difficult for you to give sustained and intensive personal attention to this project which it needed very badly." He went on to tell Gann that "We have noted with interest your preliminary evidence that the electric field can cause changes in the blood pressure and pulse of hemorrhaged animals and will see if this can be confirmed in an independent study."

There is no evidence that EPRI ever tried to replicate Gann's study. The Institute was already, however, financing a review by the Illinois Institute of Technology Research Institute (IITRI) of the scientific literature regarding the biological significance of power-line and high-voltage switchyard environments. The IITRI report of this review, which was published in 1977, started off by stating that "Research to date in Western Europe and America has failed to provide any evidence that human exposure to present levels of fields from high-voltage overhead power lines, as normally encountered, has any harmful biological effect," and that "studies performed in the U.S.S.R. report some undesirable effects on workers—not in the typical power line environment—but in the more complex environments found in high voltage switchyards." The report went on to say that "No hazardous effects were attributed to electric field exposures for laboratory and agricultural animals, as noted in some seven references from Western Europe and the United States." The seven references did not include any mention of the work of either Marino or Gann. As for magnetic-field effects, the report declared that they had been "studied extensively as a part of Project Sanguine," and that no hazardous effects had been observed. The authors went on to assure their contractors at EPRI that "Humans living in industrialized societies such as the United States have been exposed to moderate intensity, but ever increasing levels of 50/60 hertz electric and magnetic fields, generally from appliances and machinery, for nearly fifty years." They added that "This has occurred without any demonstrated cases of injury, disability, or even noticeable discomfort,'' and that ''it is very unlikely that the current levels of fields as normally encountered are doing anything very injurious to either individuals or populations.''

The only scientist in the United States to speak out against EPRI's record regarding the biological effects of electromagnetic fields from power lines was Andrew Marino. He had been in contact with Gann, and had taken the trouble to acquire a list of EPRI's research contracts. In the *amicus curiae* brief he submitted to the New York Public Service Commission in 1977, he pulled no punches about what he thought of EPRI:

> The Commission has the ultimate responsibility to ascertain that high-voltage transmission lines are operated in a manner not hazardous to human health. It might be asked, therefore, from where does the Commission expect that the facts and the information necessary to carry out its mandate will be obtained? The only organization in the United States with the capacity to provide such information is the Electric Power Research Institute (EPRI). EPRI, however, is so completely industry-oriented that all material emanating therefrom relating to health effects of high-voltage transmission lines is tainted, biased, and completely worthless.

Marino went on to describe how EPRI had canceled Gann's study ''precisely at the point where the investigators began to report serious biological effects due to ELF exposure.'' He noted that ''EPRI's Scientific Advisory Committee is heavily weighted in favor of the interests of the electric utility industry,'' and that its ''currently funded research does not reflect concern for the actual problems created by high voltage transmission lines.'' A section in his brief entitled ''Unreliability of the Electric Power Research Institute'' concluded with an analogy: ''Finally, EPRI cannot be relied upon because it makes available only a carefully selected portion of its information concerning ELF field-induced biological effects. The situation strongly parallels that described in a recent headline: 'Chocolate the World's Most Perfect Food —Report of Independent Study Group in Hershey, Pennsylvania.' ''

. . .

In January 1981, together with a copy of her letter to Roth, Wertheimer sent Kavet a letter offering to supply him and Roth with more of her data once she had finished writing the paper on magnetic fields and adult cancer. Three months later, on April 7, she wrote again, telling Kavet that Roth had called with questions about the data, and that she had supplied the answers. She added that if he or Roth wanted more information on the childhood cancer study, "this is the time when I might best be able to provide it."

Unbeknownst to Wertheimer, Roth had already finished a draft of a seven-page report entitled "An Evaluation of Health Findings in the Wertheimer Report Entitled 'Electrical Wiring Configurations and Childhood Cancer,' " and was about to submit it to Kavet. After receiving it in April of 1981, Kavet held on to it for fourteen months, and then, in June of 1982, sent it (still in draft form) to Greg Alvord, who was the scientific research coordinator for the New York State Health Department's Power Lines Project. The timing was interesting. New York health officials had just announced that they might try to replicate Wertheimer's childhood cancer study, but would wait until they had reanalyzed its data before doing so.

Roth began his report by questioning Wertheimer's conclusion that workers who were frequently exposed to alternating-current magnetic fields had developed cancer at a significantly higher rate than the population as a whole. Roth declared that this was doubtful because "the type of excess cancer deaths which occurred are those not normally associated with exposures to electric fields." As for Wertheimer's findings linking childhood cancer to high-current wiring, Roth claimed they were "almost impossible to evaluate" because they were "based on questionable statistical calculations." He went on to challenge the manner in which she had analyzed the development of cancer in relation to the amount of current that could be expected from the various types of wires in her wiring configuration code. He also criticized her for failing to take into account factors such as exposure to pollution when she selected the control population for her study. In conclusion, he asserted that because her selection of controls "might have been biased," and because her wiring configuration code "might have been incorrect," it was "doubtful that we will ever be able to properly evaluate Wertheimer's study findings."

He added, "We feel it is important for the utility industry to conduct a more comprehensive and controlled study of the Denver area and to study other geographical locations as well."

When Andrew Marino sent a copy of Roth's report to Wertheimer in August 1982, she wasted little time in firing off a four-page letter to Kavet, telling him that Roth was wrong to say that the childhood cancer study linked cancer with exposure to electric fields when in fact the link was with magnetic fields. As for Roth's suggestion that the type of excess cancer deaths she had found were not "normally associated with exposures to electric fields," Wertheimer observed that such a statement "implies that there are some types of cancer that *are* recognized to be so associated," and inquired of Kavet: "Do you know something I don't know about this?"

In responding to Roth's suggestion that her wiring configuration code "might have been incorrect," Wertheimer pointed out that he appeared to be relying on Professor Morton Miller's letter "which we have answered (and which makes mistakes a high-school physics student should be able to spot)." In conclusion, she told Kavet that she agreed fully with Roth's call for the utility industry to conduct a more comprehensive and controlled study of the Denver area and to study other geographical locations as well, and offered to help Kavet and EPRI "in any such project you may devise."

Wertheimer sent copies of her letter to Roth, Alvord, Marino, and James H. Stebbings, Jr., an epidemiologist with the Argonne National Laboratory in Argonne, Illinois, and a member of the scientific advisory panel of the New York Power Lines Project. In a cover letter to Alvord, she asked him to distribute her response to those people to whom he had sent a copy of Roth's evaluation of her work. On September 7, Kavet wrote back to Wertheimer claiming that "the Roth document you received was stamped 'DRAFT' and was intended for internal review and was not as yet intended for widespread distribution." He did not tell her why—if that were the case—he had sent it to Alvord at the New York State Department of Health.

On September 18, Wertheimer again wrote to Kavet, calling his attention to a report recently published in *The New England Journal of Medicine*, which linked the development of leukemia to occupational exposure to electric and magnetic fields.

# THE WORKPLACE CONNECTION

THE REPORT TO WHICH WERTHEIMER referred appeared in the July 22, 1982, issue of *The New England Journal of Medicine* as a letter to the editor from Dr. Samuel Milham, Jr., a physician and epidemiologist working for the Washington State Department of Social and Health Services, in Olympia. While updating an earlier study he'd done of occupational mortality, Milham examined the data for 438,000 deaths occurring among workingmen in Washington between 1950 and 1979, and he noticed that among men whose occupations required them to work in electric or magnetic fields, there were more deaths caused by leukemia than the proportionate mortality ratio (PMR) in the general population would lead one to suspect. Indeed, the PMR for leukemia was elevated in ten out of eleven occupations he investigated which were linked to electric or magnetic fields. The excess was most striking among aluminum-reduction workers, who are exposed to strong magnetic fields induced by high-amperage direct current which is used during the reduction process. These workers experienced twice as many deaths from leukemia as had been expected. Leukemia deaths were also significantly higher among electricians, power and telephone linemen, power station operators, and motion picture projectionists. (The latter work near step-up transformers that emit strong magnetic fields.) "These findings suggest that electrical and magnetic fields may cause leukemia," Milham wrote.

On November 20, 1982, *The Lancet*—the English medical journal that is one of the most highly regarded in the world—reported that Milham's findings had been supported by the results of an investigation conducted by three physicians in the Department of Family and Preventive Medicine at the University of Southern California's School of Medicine, in Los Angeles. The California researchers reviewed cases of leukemia in men from Los Angeles County who were listed as having jobs that entailed exposure to electric or magnetic fields. They discovered that the incidence of acute leukemia and acute myeloid leukemia was significantly higher than expected among these workers, especially among power and telephone linemen.

Close on the heels of the Los Angeles study came the results of an investigation into mortality from leukemia among electrical workers in England and Wales. The study was conducted by Michael E. McDowell, an epidemiologist working in the medical statistics division of the Office of Population Censuses and Surveys in London, and its results were published in *The Lancet* on January 29, 1983. McDowell analyzed 537 deaths from acute myeloid leukemia and found a consistently increased risk for the disease among all electrical occupations—the highest risk being among telecommunications engineers. An editorial in the same issue of *The Lancet* called attention to the fact that McDowell's findings followed those of Wertheimer and Leeper, Milham, and the three researchers from the University of Southern California, and observed that "the cluster of reports relating to acute myeloid leukemia is worrisome." The editorial went on to say that "Since all of us are exposed to some electric and magnetic fields, and continuously to low levels of non-ionizing electromagnetic radiation, it is important to know what risks, if any, are entailed."

During the next three years, an excess risk of leukemia was discovered among electrical workers in New Zealand, Canada, and southeast England, as well as among coal miners in the United States. (Coal miners are intimately exposed to electromagnetic fields from overhead power lines and transformers in tunnels and other closely confined areas.) Indeed, by early 1986, fifteen out of seventeen surveys of electrical and electronic workers around the world showed a link between exposure to ELF electric and magnetic fields and the development of cancer. Among them was a 1984 study showing that a significantly higher than expected number of white male residents of Maryland who had died of brain tumors between 1969 and 1982 had been employed in electrical occupations—electricians, electrical or electronics engineers, and utility servicemen. Further corroboration of this link came in 1988, when a study of men who had died of brain cancer in East Texas between 1969 and 1978 showed a significantly elevated risk for those men working in the communications, utilities, and trucking industries. Indeed, the risk for electric-utility workers was *thirteen* times that for workers who were not exposed to electromagnetic fields.

Wertheimer had first called attention to the possibility that exposure to ELF electromagnetic fields might be causing an excess of cancer in workers in certain occupations in her 1979 childhood cancer study. It was Milham's 1982 report in *The New England Journal of Medicine*, however, and its subsequent corroboration by researchers in many countries that finally began to rouse the members of the medical and scientific community out of their lethargy concerning the health hazards of ELF electric and magnetic fields—a torpor that had been induced by a decade of comforting assurances that all was well from the Navy, EPRI, the Department of Energy, and the National Academy of Sciences.

Samuel Milham is a tall, rumpled, brown-eyed, and dark-haired man of Lebanese descent who was born in Albany, New York, in 1932. He got his medical degree from Albany Medical College in 1958, and in 1961 he received his master's degree in public health from the Johns Hopkins University School of Hygiene and Public Health. He then worked in Albany for five years as an epidemiologist for the New York State Department of Health, setting up a congenital malformation surveillance system in New York State. Based on routine examination of birth and stillbirth certificates and other vital records, it was the first such system to be established in the United States, and its success was soon measured by its quick detection of congenital malformations caused by use of the drug thalidomide.

During the early 1970s, Milham, who had gone to work in the Washington State Department of Social and Health Services in 1968, analyzed more than 300,000 death certificates of male workers in the state who had died between 1950 and 1971, and coded the occupation and employer of each of them. Two major findings grew out of this analysis, which was partly financed by the Cincinnati-based National Institute for Occupational Safety and Health. One was that men who worked at a copper smelter in Tacoma owned by the American Smelting and Refining Company (ASARCO) were dying about three times more frequently of lung cancer than other workers. Presumably this was because of their exposure to arsenic—a recognized carcinogen and a by-product of the copper-smelting process. This finding led Dr. Milham and a colleague to examine children living near the ASARCO smelter, whom they found to have increased levels of arsenic in their hair and urine. Since the urinary arsenic levels of the workers at the ASARCO smelter were similar to those of children who simply lived in the vicinity of the plant, Milham and his associates warned that people in the surrounding community might be exposed to an increased risk of developing lung cancer. As a result, steps were taken to reduce arsenic emissions from the Tacoma smelter.

Another major finding was that workers employed in the half-dozen aluminum-reduction plants then operating in Washington State were dying at a significantly increased rate from cancer of the lymphatic system and the pancreas. After several major aluminum producers refused Milham's request to conduct follow-up studies of their workers, Dr. James P. Hughes, who was medical director of the Kaiser Aluminum & Chemical Corporation of Oakland, California, encouraged him to investigate mortality among workers at the company's Kaiser Mead aluminum-reduction plant in Spokane. Milham quickly found that men in this plant were dying at an increased rate from leukemia, lymphoma, Hodgkin's disease, and benign brain tumors.

"The fact that there was no great increase in lung cancer among these workers—especially those in the pot rooms—surprised and puzzled me," Milham said recently. "Aluminum pot-room workers are exposed to benzoapyrene—the culprit in cigarette smoking—because coal-tar pitch is used as a binder in the electrolytic reduction process. At the time I did my study, the prevailing wisdom in the aluminum industry blamed coal-tar pitch for the excess cancers in aluminum workers I had found. But although coal-tar pitch had long been recognized as a potent carcinogen, it had never been associated with lymphoma, so I knew that it couldn't possibly explain why the pot-room workers were dying of lymphoma at about six times the national average, and that I would have to look elsewhere for an answer to the problem.

"One thing that jumped out at me was the fact that the pot-room workers were exposed to extremely powerful magnetic fields. These fields are generated by the tremendous seventy-five-thousand-plus-ampere direct current that powers the electrolytic process, during which aluminum oxide is reduced to molten aluminum, and they are strong enough to tear a monkey wrench out of a man's hand. For this reason, I decided to look through the death certificates of all adult male deaths in Washington State between 1950 and 1979 to see if deaths from non-Hodgkin's lymphoma, as well as leukemia, might be linked to occupations with exposure to electromagnetic fields. The results of that analysis were the findings I published in *The New England Journal of Medicine* in the summer of 1982. Since then, I've learned that non-Hodgkin's lymphoma is a type of tumor that often develops in people who have impaired immune systems, so it may turn out that magnetic fields promote cancer by weakening the immune system."

Not long after his letter appeared in the *Journal*, Dr. Milham undertook to study causes of death among members of the American Radio Relay League, Inc., a national group of 140,000 amateur radio operators whose hobby exposes them to electromagnetic fields. By acquiring back issues of the League's monthly magazine and reading the obituary sections, he learned the names of 296 male members from Washington State and 1,642 male members from California who had died between 1971 and 1983. After obtaining death certificates for 280—some 95 percent —of the Washington State members, and for 1411 (about 86 percent) of the California members, he analyzed the causes of death for the combined total of 1691 according to the proportionate mortality ratio for 1976, and found excess leukemia among the radio operators.

Particularly striking was the fact that deaths from acute and chronic myeloid leukemia were nearly three times as prevalent among these men as they should have been, whereas deaths from lymphocytic leukemia were not particularly elevated. This was significant because acute and chronic myeloid leukemia are known to be caused by exposure to ionizing radiation, whereas chronic lymphatic leukemia is not as a rule caused by ionizing radiation. Also striking was the fact that 97—some 35 percent—

of the 280 death certificates from the Washington State amateur radio operators listed occupations for them such as electrician, electronics technician, and radio operator, whereas those occupations only account for about three percent of male deaths in the Washington State death file.

# PARTIAL CONFIRMATION

In September 1982, at the same time Wertheimer called Kavet's attention to Dr. Milham's study, she sent him preliminary data from a recently completed Swedish study that supported the association she had found between the development of cancer in children and the proximity of their homes to high-current wiring. The Swedish study had been conducted by Dr. Lennart Tomenius, a medical officer with the County of Stockholm. When he began his investigation in the fall of 1979, he realized that Wertheimer's results might not apply in the County of Stockholm, because most of the electricity there is transported (as it is in New York City and many metropolitan sections of the United States) in buried cables made up of closely packed and parallel distribution wires, whose current flow often tends to cancel out the magnetic fields they generate. Since Tomenius could not use a wiring configuration code in these circumstances, he decided to measure the strength of the alternating 50-hertz magnetic field outside the entrance door of each of the birth and diagnosis dwellings of young Swedish cancer victims, and to take note of any high-tension wires carrying between 6,000 and 200,000 volts, as well as any power substations, transformers, and electrified railways and subways that might be visible within 150 meters of these homes. (In Sweden, as in most European countries, electricity is generated at 50 rather than 60 hertz.) He then collected the addresses at the time of birth and diagnosis of cancer for 716 children up to the age of eighteen who were born in Stockholm County. For each of these young cancer victims he selected a control individual, who was matched according to age, sex, and church district of birth.

In the end, Tomenius surveyed a total of 2,098 dwellings of cases and controls. He found that 196 of these homes were located within 150 meters of high-tension wires, power substations, transformers, or electrified railways and subways. Another forty-five of them were within that distance of a 200,000-volt transmission line. The average strength of the alternating-current magnetic fields measured at the entrance doors of the homes near the 200,000-volt wires was 2.2 milligauss. When Tomenius analyzed his data, he found that twice as many of the homes of children who had developed cancer were near 200,000-volt lines as were the homes of control children. Similarly, children who lived in forty-eight homes with magnetic fields of three milligauss or more developed cancer twice as often as control children.

Tomenius' study was not officially published until 1986, when it appeared in the journal *Bioelectromagnetics*. However, he gave a preliminary report of his findings at the International Symposium on Occupational Health and Safety in Mining and Tunneling, which was held in Prague in June 1982. It was this presentation that Wertheimer sent to Kavet at EPRI in September. As might be expected, she was gratified by the results of the Swedish study. "It feels good not being out there alone anymore," she told Louis Slesin, founder and publisher of *Microwave News*—a monthly report about non-ionizing radiation which is published in New York—who interviewed her that autumn. Robert Becker had also seen the Swedish study, and he told Slesin it was "an important contribution in light of Wertheimer's original paper and the recent report by Milham in Washington State." Becker emphasized that the magnetic fields observed in the Swedish study were not very strong—"We are dealing with everyday exposures."

Obviously, Milham's findings and the Swedish study put pressure on officials of the New York State Department of Health and the members of the Power Lines Project's science advisory panel to replicate Wertheimer's childhood cancer study. However, in spite of the recent disclosures that tended to confirm her work, some of these officials were reluctant to concede that electromagnetic fields from power lines might constitute a health hazard. Among them was the state health commissioner, Dr. David Axelrod, who addressed the issue at the 1982 annual meeting of Central New York public health officials, held in Utica in late October.

According to an account of his talk that appeared in the Rome *Sentinel*, Axelrod ruled out a Department of Health investigation of complaints from a number of residents who lived near the Marcy-to-Massena 765,000-volt power line, and who claimed that both their own health and that of their livestock were being harmed by daily exposure to the line's electromagnetic field. The health commissioner specifically dismissed the claim of Mr. and Mrs. Richard Hoskins of Greig, New York, that an increase in abortions, stillbirths, birth defects, and infertility among the cows of their dairy herd was the fault of the transmission line, which crossed pastureland on the Hoskinses' dairy farm—operated by the Hoskins family since 1820. The Hoskinses claimed that health problems among their livestock had increased dramatically since the transmission line was energized in 1978, and for this reason they had called upon the New York State Power Authority to buy their home so that they could move away.

As it happened, Robert Becker had notified Health Department officials of the plight of the Hoskins family six months earlier, urging the department to investigate their complaints. The department officials responded that the question of whether the Marcy-to-Massena line posed a health hazard could not be answered until the research program undertaken by the Power Lines Project was completed in 1986. In the end, the Hoskinses sued the New York Power Authority, and settled out of court for $20,000.

According to the *Sentinel* account of the 1982 meeting in Utica, Axelrod declared that it was "impossible to draw any conclusions" from the fact that the Hoskinses' twelve-year-old daughter had developed a rare thyroid disease. The commissioner went on to say that his department would not direct county health officers to monitor health complaints from people living along the power-line right-of-way, on the grounds that it was the health officers' responsibility to do so anyway. A county health commissioner said that complaints about disorders among livestock would be investigated by the county's animal disease control officer, and that all his office could do was "verify any physical problems that might be described" and bring them to Axelrod. The *Sentinel* pointed out that this would not solve much since "Dr. Axelrod has already discounted the possibility that livestock ailments could be due to exposure to the high-voltage power line."

Meanwhile, Wertheimer and Leeper had finished their investigation of adult cancer and submitted a paper entitled "Adult Cancer Related to Electrical Wires Near the Home" to the *International Journal of Epidemiology*, which scheduled it for publication in December 1982. The two researchers had conducted the study of adult cancer somewhat differently from their study of childhood cancer. In addition to expanding their wiring configuration code to include new categories—such as wires carrying very high current, ordinary high current, and ordinary low current—they had measured the magnetic fields in twenty out of some two thousand homes they surveyed for the adult study, and had found these measurements to be in agreement with those predicted by their wiring configuration code.

When Wertheimer finished analyzing the data they had collected, she found that "the homes of cancer patients outnumbered control homes most clearly in the VHCC [very high current configuration] category, with the proportion of cancer homes diminishing over succeeding categories." This led her to suggest that a dose relationship "may exist between AMF [alternating-current magnetic field] exposures in this range and cancer." She pointed out that the association between wires carrying high current and cancer "is quite clear and highly significant for cancer occurring in adults before the age of 55, but is much less impressive for cancers occurring at older ages." In addition, Wertheimer found that people living in homes near wires that carried very high current developed cancer of the nervous system, uterus, and breast as well as lymphomas more readily than people living in homes near wires that carried lower current, or living far from any wires.

In the "Discussion" section of her paper, Wertheimer analyzed her observations as follows:

In this study, adult cancer was found to be associated with residence at an HCC address. Although the association was highly significant, it was considerably weaker than the similar association we observed for childhood cancer. This does not necessarily mean that AMFs affect adults less than they do children; rather, it seems likely that, in adults, the years of exposure to environmental carcinogens, as well as to AMFs from various sources, all may affect cancer rates so strongly that the effect of residential HCCs is frequently obscured.

In general, a small increment in cancer, occurring against a relatively high overall cancer rate, is difficult to demonstrate. Where the background cancer rate is low, however, that same increment will be more visible. Thus if there is a modest increase in cancer rate due to HCCs, it should be most visible in children, where cancer is generally rare; and it should also be more visible in young adults, where cancer is still relatively uncommon, than in older adults where cancer is frequent. This is essentially what we observed.

In the spring of 1982, Wertheimer sent an advance copy of the adult cancer study to Greg Alvord at the New York State Health Department. In turn, Alvord asked Dr. Charles Lawrence, director of the department's Division of Laboratories and Research, to review it. On May 17, Lawrence sent Alvord a memorandum that was highly critical of the fact that Wertheimer had failed to blind-code all of her data. While Lawrence's memo was circulating through the Department of Health in the summer and autumn of 1982, Milham published the surprising and disturbing news in the *New England Journal of Medicine* that occupational exposure to electric and magnetic fields appeared to be a cause of leukemia. At about the same time, members of the Power Lines Project's scientific advisory panel also became aware of the Swedish epidemiological investigation that partially confirmed Wertheimer's childhood cancer study.

In February 1983, Wertheimer came into possession of the memorandum that Lawrence had written nine months earlier. She then wrote to Michael Rampolla, who was the administrator of the Power Lines Project, defending her decision not to blind-code all her data and explaining that the partial blind sample she had taken had justified this decision. As it happened, Wertheimer's letter to Rampolla coincided with a decision by the members of the scientific advisory panel to commission Annemarie F. Crocetti, an epidemiologist and a clinical associate professor in the Department of Community and Preventive Medicine of the New York Medical College, in Valhalla, New York, to evaluate Wertheimer's work and determine whether it warranted further investigation by the Power Lines Project.

Crocetti, who received her doctorate in public health at Johns Hopkins, is highly regarded as an expert in the toxicity of heavy metals, such as cadmium, lead, and nickel, and she has long been active in efforts to reduce lead levels in air, soil, water, and food. In February, she traveled to Boulder and met with Wertheimer over a period of several days to discuss in detail the methodology, analyses, and results of the investigations Wertheimer and Leeper had conducted. When she returned to New York, she wrote up a 26-page report of her findings, which she submitted to the science advisory panel in April.

Crocetti told the panel that Wertheimer's data collection was "extraordinarily meticulous," but that she found the adult study to be flawed in terms of its selection of controls and in how it defined the length of time for which its subjects had been exposed to alternating-current magnetic fields. The childhood study, she felt, presented "relatively few problems," and she recommended that it be replicated as part of the Power Lines Project's research program. She also recommended that gaussmeter measurements of indoor magnetic fields be made in order to validate Wertheimer's concept of determining their strength by using wiring configuration as a surrogate, saying, "Dr. Wertheimer feels very strongly that her work was exploratory and not a full-fledged study. She has contended all along that replication is not only desirable but necessary, and she would prefer that other investigators carry out such a replication."

Shortly after Crocetti submitted her report, the members of the science advisory panel accepted her recommendation that the childhood cancer study be done again, and also decided to look into Milham's finding that adult leukemia was associated with exposure to alternating-current magnetic fields. In September, they awarded a $401,612 contract for a study of adult cancer to the Battelle Pacific Northwest Laboratories, and a $355,905 contract for a replication of the childhood cancer study to David A. Savitz, an epidemiologist at the University of Colorado's School of Medicine in Denver. In doing so, they were fully expecting (as

Dr. David O. Carpenter, director of the New York State Health Department's Wadsworth Center for Laboratories and Research, who had been placed in charge of the Power Lines Project, later admitted) that Wertheimer's findings would not hold up, and that the new studies would finally lay them to rest.

# PIONEER

NEARLY TEN YEARS HAD ELAPSED since Becker and his six colleagues on the Navy's advisory committee reviewed evidence that ELF electric and magnetic fields from Project Sanguine might adversely affect human health, and warned that a large segment of the nation's population might be exposed to even stronger fields from electric power transmission lines. During that time, dozens of experimental studies in the United States and abroad had shown that ELF electromagnetic fields produced biological effects in test animals, and the epidemiological studies by Wertheimer and Leeper, Milham, Tomenius, and others had suggested that alternating-current magnetic fields might be a cause of cancer in humans. Doubters in the medical and scientific community, who were echoed by detractors from the electric utilities industry, had challenged the epidemiological studies on the grounds that they contained insufficient data about the intensities of the magnetic fields encountered in residential and occupational exposure, and that other potential causes of cancer, such as pollution, genetics, and diet, had not received adequate consideration. What bothered the doctrinaires of the medical and scientific establishment most of all, however, was the lack of theoretical ground upon which to support the hypothesis that the weak electric and magnetic fields generated by power lines could interact with biological systems, let alone pose a threat to human health.

As it happened, information that would help to explain the observed effects of power-line radiation had been in the process of being developed since the early 1970s, thanks to the pioneering efforts of an Australian-born scientist, Dr. W. Ross Adey, who had served on the National Academy of Sciences' Committee that evaluated Seafarer. Together with colleagues at the Brain Research Institute of the University of California at Los Angeles, Adey had been investigating the interaction of weak electromagnetic fields with the outer membranes and inner workings of brain cells. Adey is a lanky, brown-eyed man of sixty-seven with wispy gray hair that falls over his ears, and a craggy, weather-beaten face. He runs regularly in fifteen-mile road races and spends weekends climbing shale faces in the San Bernardino Mountains. He has a wry sense of humor, a combative streak, and a marked inability to suffer fools gladly—indeed, a strong tendency not to suffer them at all—and he has not only survived the controversy, uncertainty, cover-up, and political infighting that for two decades have characterized the field in which he works, but has also emerged as its leading theoretician and philosopher.

Adey was born and raised in Adelaide, in South Australia. Like many children of the twenties, he was fascinated by the radio, and as a teenager he built a large radio receiver which he used to record signals from the planet Jupiter. He also became a ham radio operator, and remains one to this day. In 1939, he entered the University of Adelaide Medical School, and after receiving his bachelor's degree in medicine in 1943, he served for two years as a surgeon lieutenant in the Australian Navy, where he became familiar with the techniques of radar. Following his service in the Navy, he went back to medical school in Adelaide and received his doctorate in 1949. He had intended to be a neurosurgeon, but the subject of his degree thesis caused him to change his mind and he chose instead to be a clinical neurologist and neuroscientist.

"I did my thesis on the nervous system of a giant five-foot-long earthworm called the Megascolex, which is found in the hills above Adelaide," he said recently. "I found that the tiny, rudimentary brain of this creature has only two dozen or so nerve cells, sitting shoulder to shoulder, and making contact with each other not by way of connecting fibers but with dendrites—branchlike projections that sprout from the nerve cells and allow them to communicate with each other by sending slow wave impulses. What struck me as most significant was that this dendritic apparatus provided the primitive brain of the Megascolex with a powerful means of communication. In short, it enabled this creature to have a repertoire of complex behaviors far greater than one would expect from so few nerve cells.''

After receiving his doctorate, Adey went to England and spent the next two years as an Oxford Research Fellow at Oxford University, where he studied anatomy and the physiology of the limbic system—the portion of the brain that is concerned with the expression of emotional and sexual behavior, and is also associated with memory. In 1951, he received a Rockefeller Grant to study in the United States, and while touring in California, he met Horace W. Magoun, a world-renowned brain physiologist from Santa Monica who had founded the Department of Anatomy at the University of Southern California at Los Angeles, and later founded the Brain Research Institute at UCLA. At Magoun's invitation, Adey came to the United States from Adelaide in 1954 and went to work as a professor of anatomy and physiology in the Department of Anatomy at UCLA, and in 1961 he joined the university's newly formed Brain Research Institute.

During the early 1960s, Adey did groundbreaking work on the radio transmission of brain waves, placing tiny transmitters in the skulls of test animals and broadcasting their electroencephalographic (EEG) waves to receivers. In 1965, he became director of the newly established Space Biology Laboratory at the Brain Research Institute, and over the next two years he and his colleagues performed extensive computer analyses of the EEG patterns of fifty astronaut candidates from the National Aeronautics and Space Administration.

A few years earlier, Rafael Elul, an Israeli scientist who had taken a leave of absence from the Hebrew University-Hadassah Medical School in Jerusalem to work in Adey's lab, had performed a series of pioneering experiments in which he inserted microscopic electrodes into the tiny, fluid-filled spaces between the cells of cultured cat-brain tissue. He also penetrated the nerve-cell membrane and was among the first to record the unique rhythmic waves that were generated inside cerebral nerve cells.

Elul found that the spaces between the cells contained sugar-tipped strands of protein that protruded from the membranous surfaces of the cells. In addition, he learned that the tips of these protein strands were extremely sensitive to positive or negative currents, and could be seen to sway in response to them.

"At that time, brain waves were widely regarded as being a general 'noise' in the cerebral system, and as having little or no direct physiological role in information processing," Adey recalls. "It was also believed that the excitation of nerve cells required powerful electric stimulation, in order to overcome the tenth-of-a-volt electrical field that is known to be present at all times between the exterior and the interior of every cell membrane. There appeared to be good reason for this, because the electric charge between the exterior and interior of the very thin cell membranes, which is known as the membrane potential, is enormous. It is, in fact, equal to the charge of a two-hundred-thousand-volt power line, if, instead of being suspended fifty feet in the air as is customary, the power line were to be placed an inch above the ground. In such a case, though, the power line would, of course, instantly arc over and burn out. That gives some idea of the astonishing resistance to electrical current that is provided by the two layers of fat—only a millionth of an inch thick—that make up the membrane of a cell. Indeed, the insulating quality of the tiny cell membrane is vastly more efficient than most, if not all, man-made materials. The wonder is that every cell in the body functions from birth to death with this extraordinary electrical barrier at its surface. A major question to be answered, therefore, was how could this be so?"

Adey remembers that by the early 1960s he had begun to wonder if cell membranes might not play a much more important role in brain function than anyone suspected. "Then Elul came along and provided us with a breakthrough," he says. "Elul gave us a new appreciation of the protein arborization protruding from the cells, and showed us that these strands might act as Trojan horses to permit extremely weak electrical and chemical signals to pass through the barrier of the membrane potential and reach the cell interior. His work paved the way for us to consider the possibility that the protein strands might be sensitive to electrochemical breezes blowing across the cell membranes—much in the manner of a field of wheat waving in the wind—and to hypothesize that the rhythmic waves they were generating were not just general noise but intercellular whispering—in other words, the sound of brain cells communicating with each other in a private language. Since all brain waves, whether they are dominant EEG waves or the weak electric ripples emanating from the interiors of cells, travel as oscillations through the fluid-filled spaces—or gutters, as I prefer to call them—it was obviously important for us to know more about what was going on in these gutters. It was also necessary for us to learn more about the role of calcium, whose prevalence in brain tissue had fascinated me since my days as a medical student, and whose ions were known to play important roles in the transmission of nerve impulses in the brain.''

In the mid-1960s, Ahron Katchalsky, a Russian-born molecular biologist and biophysicist from Israel who was a member of the Massachusetts Institute of Technology's Neurosciences Research Program, began to do just that. Katchalsky investigated the binding of calcium, magnesium, hydrogen, and other important biological ions to artificial molecules that simulated the protein strands found in all living tissue. In the course of his experiments, he discovered that calcium ions are bound more strongly to the protein strands than are most of the other ions present in the fluid-filled gutters of the brain. At about the same time, other scientists demonstrated that calcium might be involved in chemical changes taking place at the membrane surface. At that point, it became clear to Adey that if he was going to learn anything about the role of the cell membrane in brain chemistry, he was going to have to investigate the binding and unbinding of calcium. So when the space program began to wind down in 1970, he decided to pursue this line of inquiry in earnest.

As it happened, the approach used by Adey and his colleagues in studying cerebral calcium was influenced by a study they had conducted for the government in 1967 and 1968, which, in turn, resulted from the 1962 discovery by the State Department that the Russians were beaming low-intensity microwave radiation into the American Embassy in Moscow. At first, it was believed that the microwaves were being directed into the embassy to activate electronic listening devices that had been hidden in its walls. It was soon realized, however, that the Russians were transmitting at multiple frequencies that did not appear suitable for normal electronic eavesdropping techniques, and by 1965 the motive for the Soviet microwave bombardment had become the subject of intense scrutiny by various American intelligence agencies, including the Central Intelligence Agency, whose officials suspected that the Russians were conducting research on the effects of microwaves upon human behavior.

For security reasons, this scrutiny was carried on with the utmost secrecy, and cloaked in the usual euphemisms. The microwave beams being directed at the embassy were referred to as the Moscow Signal; the investigation of them was carried out under a highly classified research program, called Project Pandora; and information about it was parceled out on a strict "need to know" basis—which, as it turned out, excluded most of the State Department employees at the embassy who were being irradiated. Some idea of what was afoot can, however, be gleaned from the fact that CIA agents interviewed a number of scientists involved in microwave research, asking them such questions as whether it was reasonable to believe that microwaves beamed at human beings from a distance could affect the brain and alter behavior.

Project Pandora was placed under the direction of the Advanced Research Projects Agency (ARPA)—a secret organization within the Department of Defense, which was then engaged in developing a wide variety of electromagnetic weaponry. In 1966, ARPA set up a special laboratory at the Walter Reed Army Institute of Research in Washington, D.C. Over the next two years, monkeys were irradiated with microwaves at power intensities and frequencies similar to those of the Moscow Signal, which were said to be in the gigahertz range (in other words, billions of cycles per second), in order to determine whether the radiation could induce biological and behavioral changes. At the same time, ARPA gave a number of research contracts to independent scientists to study the effects of low-level radiation in test animals. Among these was a contract awarded to Adey and some colleagues at the Space Biology Laboratory to investigate the effect of weak ELF electric fields oscillating at seven and ten hertz on the brain waves and behavior of monkeys.

Toward the end of 1968, a special five-member science advisory committee issued a report stating that monkeys exposed to the simulated Moscow Signal by Dr. Joseph C. Sharp, director of Walter Reed's electromagnetic radiation laboratory, "showed degradation of work performance." During the winter of 1969, ARPA and the Navy conducted a pilot study to determine the physiological and psychological effects of microwave exposure from radar upon twenty-one crew members of the aircraft carrier USS *Saratoga*, without bothering to inform the crew members that they were the subjects of a biological experiment. And in the spring of 1969, the special advisory committee urged that Walter Reed develop a full-fledged program of human experimentation with the Moscow Signal. The motive for the new program became clear in June, when Richard S. Cesaro of ARPA, the overall director of Project Pandora, declared that there was evidence that low-level radiation could penetrate the central nervous system of monkeys, and that it was now necessary to determine "whether the Soviets have special insight into the effects and use of athermal radiation on man."

During the next few months, however, a mysterious change of mind took place among the people who were carrying out and evaluating the Project Pandora research program. In August, Major James T. McIlwain, a research scientist at Walter Reed who had previously worked in Adey's laboratory, informed the advisory committee that the conclusions drawn from the monkey experiments conducted by Dr. Sharp were erroneous. In January 1970, after repeated and unsuccessful attempts were made to resolve the differences between the two men, the members of the committee reversed their earlier opinion, declaring that "No definitive answer to the question of whether the original signal has any effect on the performance of operantly-conditioned monkeys has been provided to date," and that "the findings thus far can be regarded as negative."

Some observers believe that the committee's revised opinion was engineered by higher authority, pointing to the fact that McIlwain's criticism of Sharp's work was never put into writing, and suggesting that the government may have decided that any plan to perform human experiments with the Moscow Signal would surely be leaked to the public if it were not conducted in total secrecy. Be that as it may, no one had anything critical to say about the experiments that were carried out for ARPA by Adey and his colleagues at the Space Biology Laboratory, because they produced incontrovertible evidence that low-level, low-frequency electric fields could affect the performance and alter the brain waves of operantly trained monkeys.

Together with Rochelle J. Gavalas-Medici, an experimental psychologist, and other associates, Adey conducted a series of experiments during which three monkeys in whose brains electrodes had been implanted were trained to press a panel once every five seconds; if they performed within a two-and-a-half-second interval, they were rewarded with apple juice. After the animals were performing satisfactorily, they were exposed for four hours a day to very weak ten-volt-per-meter electric fields oscillating at either seven or ten hertz. Exposure to ten-hertz fields did not produce any marked change in the monkeys' behavior. But when the animals were irradiated with seven-hertz fields, their ability to estimate a five-second interval was shortened by as much as half a second. Moreover, at both the seven- and ten-hertz frequencies, Adey and his co-workers observed increases in EEG intensity in the limbic system of the brain in all three monkeys, indicating that externally applied electric fields could drive the electrical activity of the brain and thus produce a phenomenon called brain entrainment, which had obvious application for mind control.

During the early 1970s, Adey, Gavalas-Medici, and Suzanne M. Bawin, a Belgian-born research neurophysiologist who was doing her Ph.D. work under Gavalas-Medici's supervision, studied the effects of low-intensity electric fields of 7, 45, 60, and 75 hertz on the brain waves and behavior of monkeys. Their one consistent finding was that weak fields (some as low as one volt per meter) oscillating at brain-wave frequencies could produce significant changes in behavior and in the EEG. This finding was reinforced when they exposed live cats to brain-wave-intensity levels of very-high-frequency (VHF) radiowaves, oscillating at 147 million hertz (147 MHz)—a frequency used by ham radio operators—modulated at brain-wave frequencies. What they did, in effect, was to impose a slow rhythm on a fast oscillation. They discovered that unmodulated 147-MHz radiation at a field intensity of one milliwatt per square centimeter had no biological effect, but that an ELF-modulated signal of the same intensity could powerfully change the pattern of the cats' brain waves at specific sites within the brain and at specific ELF frequencies. Among other things, this indicated to Adey and his colleagues that radio-frequency fields at power levels far below those necessary to induce the heating of tissue could produce profound changes in the electrical activity of the brain.

In 1974, Bawin, Adey, and Leonard K. Kaczmarek, a newly graduated biochemist from the Charing Cross and Westminster

Hospital Medical School in London (he is now a professor of biochemistry at Yale), exposed freshly removed chick brains to the same low-power, ELF-modulated 147-MHz field that had been used in the cat experiments, and found that calcium outflow was increased by about twenty percent as a result of exposure to specific frequencies between six and twenty hertz, with a maximum effect at sixteen hertz. A graph of the results showed that the outflow of calcium followed a "tuning curve," with the peak of the effect at sixteen hertz and less effect at both higher and lower frequencies. (By way of contrast, the unmodulated carrier wave of 147 MHz produced no effect.) Thus, Bawin, Kaczmarek, and Adey concluded that the phenomenon did not depend simply upon the intensity of the electromagnetic field, but was sensitive to a narrow "window" of frequencies. In later work, Adey and his associates would observe similar windowing with respect to a narrow range of field intensities and to lengths of time of exposure.

By the end of 1975, Adey and his co-workers had gathered considerable evidence to show that weak electromagnetic fields have a direct effect on the vertebrate nervous system. Their monkey studies had produced behavioral effects indicating that weak oscillating fields could result in modification of recall and the ability to estimate time. Their experiments with chick-brain tissue and awake cats indicated that weak fields could alter the flow of calcium from the membraneous surfaces of cells and from their protruding glycoprotein strands, thus changing the chemistry of the brain. These findings led them to conclude that slow electrical oscillations within the forest of protein strands surrounding brain cells formed the basis of an independent cell-to-cell communications system, and to speculate that such oscillations might be the basis for changes in brain chemistry associated with the storage of information.

What remained unclear was how electromagnetic fields that were less than one four-thousandth the strength of the electric barrier provided by the membrane potential could act as an electrical stimulus capable of influencing the release of calcium and other chemicals from the surfaces of brain cells. In an attempt to explain the mystery, Adey pointed out that, the membrane potential notwithstanding, a brain cell is so sensitive to fields that surround it that if one terminal of a 1.5-volt battery were connected to a wire placed in the Pacific Ocean at San Diego, and the other terminal to a wire in the Pacific Ocean near Seattle—about a thousand miles away—the cell would be able to detect the infinitesimal electric gradient of the resulting one ten-millionth of a volt per centimeter that would be produced in the intervening seawater. Adey then postulated that weak electromagnetic fields must be acting as triggers to stimulate a powerful system of cooperative interactions at the membrane surface. As for why these electromagnetic fields did not elicit major disruptions of cerebral processes, such as producing perceptual distortion or interfering with judgment and decision making, Adey suggested that, because each cell in central nervous tissue generates its own very small electrical field within its own immediate and tiny environment, an electrical field originating outside the nervous system would be far too crude to disturb the complex, intrinsic field within. In a paper written at the end of 1976, he drew an analogy with ocean waves: "Within a cubic meter of water at the ocean surface there are numerous local pressure gradients producing highly complex focal motions. Passage of a large tidal wave through the same region has relatively little effect on the major fluid movements within this small volume."

Up to this point, Adey's research at the Space Biology Laboratory had been financed almost entirely by NASA, ARPA, the National Science Foundation, the Air Force Office of Aerospace Research, and the Office of Naval Research. This circumstance undoubtedly influenced his decision to become a member of the National Academy of Sciences' committee formed in January 1976 to evaluate the possible biological effects of the ELF electric and magnetic fields that would be generated by the Navy's Project Seafarer. At that time, he was not yet prepared to make a major issue of his opposition to the thermal theory of microwave and radio-frequency injury, which represented the mainstream thinking of the scientific community on the biological effects of non-ionizing radiation. Some observers note that Adey and his colleagues were just beginning to probe the surface of something profoundly revolutionary in the realm of science, and realized that there were many people in positions of authority and responsibility who might be capable of hindering the painstaking program of research they were about to undertake.

Whatever the case, when Robert Simpson, the attorney for the New York Public Service Commission, who had recruited Becker and Marino, asked Adey to testify about the biological effects of ELF fields at the power-line hearings, he replied that the issue was far too complicated for lawyers to handle, and that he wanted no part of it. He evinced a similar attitude when approached by a journalist at a scientific meeting in Boulder, Colorado, in 1975, expressing doubt that lay readers of the press would be able to grasp the difference between biological effects and biological hazards, and voicing concern that the great potential of radio-frequency radiation as a tool for studying the electrochemical workings of the brain might go undeveloped if there were too much adverse publicity. Over the next ten years, however, Adey would change his mind, as he came to realize that the ignorance of laymen and the possibility of adverse publicity posed nowhere near as much of a risk to the search for the truth about ELF electromagnetic fields as did the obfuscation of industry, the mendacity of the military, and the corruption of ethics that industrial and military money could purchase from various members of the medical and scientific community.

# PART TWO

---

# Cover-up

Chapter

# "WHO IS MY ENEMY?"

IN THE SPRING OF 1977, Ross Adey left UCLA after 23 years at the Brain Research Institute to become associate chief of staff for research and development at the Jerry L. Pettis Memorial Veterans' Hospital, in Loma Linda—some 80 miles east of Los Angeles—and professor of physiology at the Loma Linda University School of Medicine. "I left UCLA because I wanted to pursue lines of research in an environment free of establishment views regarding cell and molecular biology," he said recently. "Linear equilibrium thermodynamics—in this context, the transfer of energy by heating—has been a cornerstone of biology and medicine for the past three hundred years. It is of very limited utility, however, in understanding the essence of living matter and, in particular, of the essential steps in the coupling of signals across cell membranes. There, as in all other biomolecular processes, we have come to the awesome realization that the real story is written at the atomic level in physical terms, rather than in chemical terms in the fabric of molecules."

By the time he went to Loma Linda, Adey and Suzanne Bawin had reported extensively on their studies showing that weak ELF fields of between 1 and 100 hertz, and also ELF-modulated VHF fields of 147 MHz, could alter the outflow of calcium ions from the isolated brain tissue of chickens, as well as from the brain tissue of living cats. With financial support from the Food and Drug Administration's Bureau of Radiological Health, the two colleagues, along with Albert F. Lawrence, an applied physicist in Adey's lab, had also demonstrated that the same chemical changes could be effected by exposing the brain tissue of living cats to low-intensity microwave fields of 450 MHz, providing that the carrier frequency was modulated at the ELF frequency of 16 hertz.

As it happened, this was extremely close to the frequency of a new type of radar that the Air Force was planning to install at the Massachusetts Military Reservation on Upper Cape Cod—the Otis Air National Guard Base is part of the reservation—and at Beale Air Force Base, some 35 miles north of Sacramento, California. The new radar was called PAVE PAWS—an acronym for Precision Acquisition of Vehicle Entry Phased Array Warning System. Unlike conventional radar with its rotating antenna, PAVE PAWS was equipped with an array of more than 10,000 solid state components, called radiating elements, which could be individually controlled by computers to form a single beam that could be steered electronically and very rapidly in a large number of directions over a 240-degree field. Designed to detect sea-launched ballistic missiles, PAVE PAWS was powerful enough to resolve an object the size of a basketball at a distance of 1500 miles. It operated at a carrier frequency of between 420 and 450 MHz, with a pulse repetition rate of 18.5 hertz—very close to the 16-hertz modulation frequency of the 450 MHz radiation that Adey, Bawin, and Lawrence had used to change the brain chemistry of living cats.

As things turned out, this similarity would place Adey and the Air Force on a collision course. In March 1976, the Air Force had issued a 250-page environmental assessment, which, except for a brief mention of the fact that emissions from PAVE PAWS might cause cardiac pacemakers to skip a beat or two, ignored the possibility that the radiation levels emanating from PAVE PAWS might prove hazardous to human health. The Air Force assessment was, however, full of assurances that the radar would not harm the water table, the white-tailed deer, the wood duck, or the Wilson's petrel. Indeed, the document appeared to have been written with the notion that the environment was something inhabited exclusively by birds and animals.

An assertion by the chief of the Air Force's Environmental Policy and Assessment Branch that PAVE PAWS would not have "a significant effect on the quality of human environment" furnished the Air Force with a convenient excuse for not preparing an environmental impact statement, which, unlike an environmental assessment, must be submitted to the Environmental Protection Agency, circulated to federal agencies and the public for comment, and be rewritten to respond to those comments. In this way, the Air Force hoped to avoid the prospect of holding public hearings on PAVE PAWS and of having to answer questions about it.

During 1976, the Raytheon Company—the prime contractor for the PAVE PAWS system—was hard at work making the radar a reality on Cape Cod. As a result, long before Cape Codders became aware that radiation from PAVE PAWS might prove hazardous to their health, the radar edifice—a massive 105-foot-tall truncated pyramid of reinforced concrete which had been designed to withstand nuclear blast—loomed above the horizon only 3,500 feet from the heavily traveled mid-Cape highway. By contrast, in California, where construction of PAVE PAWS did not begin until April 1977, the Air Force was quickly forced to answer questions about the safety of the radar as a result of local newspaper articles that disputed some of the claims made in the environmental assessment.

The way in which the Air Force responded to these inquiries clearly revealed how anxious it was to allay any adverse public reaction to PAVE PAWS. On April 22, 1977, Lieutenant Colonel Paul T. McEachern, director of the PAVE PAWS program, told a group of Californians that there would be no health hazard from PAVE PAWS. At the same time, he assured them that since the Air Force would have had over a year's experience with the PAVE PAWS on Cape Cod, "we'll have a fairly good idea of the effects on the environment before we even turn it on here."

McEachern's attempt to assuage the Californians occurred slightly more than two weeks after Adey had described in detail the ability of low-intensity 450-MHz radiation modulated at 16 hertz—very similar to the radiation pattern that would emanate from PAVE PAWS—to alter the brain chemistry of chicks and cats. Adey discussed his findings at a meeting of the Electromagnetic Radiation Management Advisory Council (ERMAC) held in Washington, D.C., on April 5. Among the participants at the meeting was John C. Mitchell, the chief of the Radiation Sciences Division of the Air Force's School of Aerospace Medicine in San Antonio, Texas. Two months earlier, Mitchell had been present at another meeting of ERMAC at which Adey had declared that research being conducted at the School of Aerospace Medicine "lacked relevance to the study of normal brain function." Adey had also questioned why the Air Force was conducting most of its brain experiments with continuous electromagnetic fields, rather than the more biologically potent, pulsed radiation that was being given off by all of its radars.

As might be expected, Cape Codders were taken aback some months later when they learned that Lieutenant Colonel McEachern had, in effect, described them as test animals for whatever the long-term chronic effects of PAVE PAWS radiation might be. They soon communicated their concern to their representative, Gerry E. Studds, a Democrat from the Twelfth District of Massachusetts. On January 9, 1978, Studds asked the Air Force to document the research it had used to determine that long-term exposure to low-level microwave radiation would not endanger human health. "If the Air Force cannot justify the PAVE PAWS construction with this type of information, then I will request that all testing of the PAVE PAWS installation be postponed until the Air Force adequately addresses this problem in a fully documented Environmental Impact Statement," he said. Studds added that the present-day actions of the nation's military and business interests must not be allowed to endanger the public health twenty years from now. "The burden of proof clearly lies on their shoulders," he declared.

As things turned out, this burden proved somewhat heavy for Mitchell and McEachern, who appeared that same week at a three-hour public forum in the auditorium of the Wing Elementary School in Sandwich, a town whose center lies less than two miles northeast of the PAVE PAWS radar. For most of the meeting, Mitchell, who was appearing as a representative of the surgeon general of the Air Force, maintained that the Air Force had taken a "conservative approach" toward the possible biological hazards of microwave radiation. However, when a young woman in the audience asked him if he could guarantee the safety of her two children from the effects of radiation from PAVE PAWS, he said that he was not going to guarantee her anything. "Then I ask you," she countered, "who is my enemy?"

By now, it was obvious that the Air Force had made a serious miscalculation in its handling of PAVE PAWS. The Air Force had figured it could stick the radar on the best available piece of real estate, tell the locals it would be good for business, assure the Audubon Society that it would be okay for birds, and let everybody think it would be harmlessly scanning the skies. At the end of January, however, the environmental groups that had sponsored the meeting had formed the Cape Cod Environmental Coalition, Inc., and retained a lawyer. In March, the Coalition filed suit in Federal District Court in Boston against Air Force Secretary John C. Stetson and several other Air Force officials, charging that the Air Force had violated the National Environmental Policy Act of 1969 by failing to submit an environmental impact statement for the PAVE PAWS project.

Some Cape Codders had begun to suspect that the Air Force might be trying to hide more than information about the biological effects of low-level radiation. Good reason for thinking so came from California, where residents of Yuba County had filed suit in August 1977 to halt construction of the PAVE PAWS at Beale Air Force Base. Just prior to a scheduled hearing in Federal District Court in Sacramento, the Air Force had offered to settle the lawsuit by guaranteeing that emissions from the radar would not exceed certain levels. (It seemed obvious that the Air Force wanted to avoid a court fight for the same reason it wanted to avoid writing an environmental impact statement; either way it would be forced to answer questions about the microwave hazard for which it had no answers.) However, negotiations for a settlement broke down when the Air Force refused to agree to the plaintiffs' demand that radiation levels from the California PAVE PAWS be continuously monitored.

At that point, Senator Alan Cranston of California wrote to Secretary Stetson asking why the Air Force had refused to agree to continuous monitoring. The Air Force Secretary replied that continuous monitoring would constitute an unnecessary expense, but a far more likely reason was that the Air Force had been something less than candid about the elevation angle of the main beam of PAVE PAWS. In the environmental assessment it had issued in March 1976, the Air Force maintained that the main beam would not operate below three degrees above the horizon. The assessment went on to say that the beam "will never be allowed to radiate along the ground or to illuminate obstructions." During the public meetings in North Truro and Sandwich, McEachern and other Air Force representatives reiterated this claim on several occasions, leaving the clear impression that the main beam of the radar would be pulsing far above everyone's head.

Whether the Air Force should have been believed in this matter was even then open to question. The Raytheon Company's contract specifications for PAVE PAWS, which are dated September 3, 1976—nearly six months after the environmental assessment was issued—stated that the main beam of the radar could be adjusted within eight seconds "to any operator-selected configuration from zero degrees (horizon) to ten degrees in elevation . . ." In fact, not until September 1977, when it was faced with the prospect of going to court, did the Air Force modify the contract with Raytheon and limit the elevation angle of the beam to three degrees. This, of course, raised the question of whether the modification of the contract was genuine or simply a ruse to conceal the Air Force's intention to maintain the option to lower the beam to zero degrees in the event this should be deemed necessary. (A case for the latter possibility can be found in several Department of Defense manuals that describe the advantages of "ground grazing" by radar beams for the early, low-altitude detection of sea-launched ballistic missiles—the acknowledged mission of PAVE PAWS.) In any event, if the Air Force did intend to operate the main beam at zero degrees, it could not very well do so in secrecy if it agreed to continuous monitoring of the radiation emanating from PAVE PAWS.

In April 1978, the Air Force announced that it had awarded a several-hundred-thousand-dollar contract to SRI International (formerly the Stanford Research Institute)—a consulting firm in Menlo Park, California—to prepare an environmental impact statement for the radar. In August, Mitchell and an Air Force survey team, along with civilian monitors, set out to measure microwave radiation levels emanating from the partly completed PAVE PAWS radar. When the preliminary results of their survey were discussed at an ERMAC meeting in Washington, D.C., at the end of the month, it came to light that their measurements of

PAVE PAWS radiation levels were about one thousand times lower than the calculations of expected power levels that had been made by the EPA. This led to speculation that summer foliage may have attenuated the radiation and affected the accuracy of the Air Force measurements. The trouble with the Air Force field survey, however, went far beyond foliage. The trouble was that the Air Force survey team had diluted its measurements by time-averaging the power density of the microwave radiation. The team did this by including in its calculations the intervals between the radar's short pulses, with the result that the average power calculated was only a small fraction of the peak pulse power. This expedient allowed the Air Force to say that it was adhering to the recommended standard for exposure to microwave radiation, of 10 milliwatts per square centimeter, which was based on the obsolete notion that its only harmful effect was the heat it could create in the body if it was generated at high power. Thus the Air Force was able to ignore mounting evidence that the instantaneous, non-thermal effect of being bombarded several hundred million times a second by powerful peak pulses of radiation which penetrate deeply into tissue can have serious biological consequences.

By this time, Representative Studds, Senator Edward M. Kennedy, and Senator Edward W. Brooke III—all of whom had previously written strongly worded letters to the Secretary of the Air Force expressing concern about the potential health hazards of PAVE PAWS and demanding that the Air Force produce an environmental impact statement for the radar—had begun to distance themselves from the controversy. The stands taken earlier by these three politicians had been prompted by letters, telegrams, and telephone calls of inquiry and alarm from their constituents. By the late summer of 1978, however, their mail and messages regarding the radar had dwindled, and there was little political mileage in their continuing to criticize the conduct of the Air Force, especially since the Air Force had agreed to produce an environmental impact statement.

The politicians had correctly gauged the temper of their constituents. Further evidence that the will to fight against the radar had begun to diminish came in November 1978, when the Cape Cod Environmental Coalition agreed to suspend its lawsuit in return for the opportunity to participate in further environmental studies of the impact of PAVE PAWS—an almost meaningless

concession on the part of the Air Force, which had already commissioned and was paying for environmental studies with several
hundred thousand dollars of the taxpayers' money. Under the
agreement, the Air Force was allowed to continue its construction of the radar, and the citizens' group was given twenty-one
days to amend its legal complaint once the Air Force filed a final
environmental impact statement with the EPA. If, within this
period of time, the Coalition failed to raise claims that the statement was insufficient, the court action would be dismissed.

Chapter

# 14

# ANSWERING THE QUESTION

MEANWHILE, BACK IN MAY 1978, the Air Force had taken a leaf out of the Navy's book and commissioned the National Academy of Sciences (also, of course, with taxpayers' money) to determine whether preliminary measurements to be made of PAVE PAWS radiation could accurately predict the extent to which people living in towns surrounding the radar would be irradiated, and whether their exposure to the radiation emitted by PAVE PAWS would constitute a health hazard. The Academy's National Research Council established two separate committees to answer these questions—a seven-man engineering panel to evaluate the amount of radiation that would be generated by PAVE PAWS, and an eight-man panel to evaluate the biological effects of the radar system. Among the members of the bioeffects committee were Brockway McMillan, vice president in charge of military systems for the Bell Telephone Laboratories—a company that had been instrumental in setting up the ten-milliwatt standard in the late 1950s; George M. Wilkening, the director of environmental health at the Bell Labs, who had sat on the Academy committee that sanctioned Project Seafarer; Dr. Don R. Justesen, director of the Neurophysiology Research Laboratory at the U.S.

Veterans Administration Hospital in Kansas City, Missouri, who, a year later, would testify in behalf of the New York Telephone Company that exposure to low-level radiation from a microwave-relay tower that the telephone company proposed to construct in Chappaqua, New York, was no more hazardous than the heat exchanged by two people kissing; and Ross Adey.

As Adey and his fellow committee members began the task of assessing the medical and scientific literature on the biological effects of microwaves, a sixteen-member team at SRI International, which had been doing the same thing, was putting together a 400-page draft environmental impact statement for the Air Force. The draft statement, which SRI submitted to the Air Force in November, and the Air Force filed with the EPA on December 22, 1978, stated that the preponderance of biological effects was found at power densities in excess of 2,000 microwatts per square centimeter, with a few scattered effects of "no medical significance" occurring at levels down to about 500 microwatts. The fact was that because the military and the electronics industry had been allowed to maintain financial control over microwave research for two decades, most studies had been designed to look only at the heating effects of microwaves. Among the relatively few low-level microwave studies that had been conducted, however, some showed that changes in the bio-electric functioning of the brain, in brain chemistry, and in the immune system could occur at levels far below those capable of producing heat.

The authors of the draft statement went on to cite as inaccurate a 1962 study showing that mice irradiated with microwaves generated by an Air Force radar transmitter had developed leukemia three and a half times more often than non-exposed control animals. "There is no evidence that the radio-frequency radiation caused leukemia in the mice," they stated. The study in question had been financed by the Air Force and conducted by Professor Charles Susskind, a member of the Department of Electrical Engineering at the University of California at Berkeley. Together with a colleague, Susskind had exposed two hundred male mice to X-band-frequency radiation that was generated by an Air Force radar transmitter. The mice were irradiated at a power density of 100 milliwatts per square centimeter for four and a half minutes a day, over a period of 59 weeks. The longevity of the mice did not appear to be affected by these conditions. However, when Susskind and his colleague performed autopsies on irradiated test animals that had died during the period of the experiment, they found cancer of the white blood cells—lymphatic leukosis and lymphatic leukemia—in fully 35 percent of them. By comparison, autopsies performed on control animals that had not been exposed to microwaves revealed cancer in only 10 percent. Interestingly, the Air Force refused to provide additional funds to repeat this study, and SRI's draft impact statement omitted any reference to the fact that officials from the Food and Drug Administration's Bureau of Radiological Health had described its outcome as "the most discomfiting finding in the available literature."

The authors of the draft statement could hardly fail to mention the results of a study in which researchers at SRI International had exposed 33 pregnant squirrel monkeys to levels of microwave radiation equal to those of the recommended governmental standard, and then exposed the offspring of eighteen of these monkeys for up to six months after birth: four out of five of the irradiated infant monkeys had died as compared with no deaths in a group of non-irradiated control animals. They claimed, however, that the study was only marginally significant because of the small number of animals involved. Nowhere in the statement did they mention a study performed by scientists at the Southern Research Institute in Birmingham, Alabama, which revealed an excessive fetal death rate among infants born to women at the regional hospital at Eglin Air Force Base in Florida. Nor did they mention that these scientists had concluded that the results of this study provided "additional evidence that a health problem may be associated with radar." The more than marginal significance of this observation was that Eglin Air Force Base was the one installation in the nation that had housed an operational PAVE PAWS–type phased-array radar over an extended period of time. In fact, the phased-array radar at Eglin had been operating since 1965.

And what of the extraordinary finding of Adey and Bawin that low-level radiation of the PAVE PAWS frequency of 450 MHz, when rhythmically modulated at a repetition rate of 16 hertz (close to the 18.5-hertz repetition rate of the PAVE PAWS radar), had produced changes in the chemistry of chick brains? The authors of the draft statement called this effect "curious and interesting," before noting that it had been observed in chick brains that had been "removed from the animals and placed in incubation dishes." They then declared that no observations existed on brains that had not been removed from the skull, and that "the significance of the finding for human health was unclear." They were in error; Adey and Bawin had observed changes in calcium outflow in the cerebral cortex of living cats, and the significance of this finding for human health would remain unclear until studies were conducted of human beings who had been exposed to radiation emanating from PAVE PAWS to determine if they had undergone similar changes in brain chemistry.

On the evening of January 22, 1979, several hundred Cape Cod residents and observers attended a public hearing convened by the Air Force in the auditorium of the Sandwich Junior-Senior High School to discuss the draft environmental impact statement. At the meeting, which was conducted by a military judge, Mitchell declared that measurements at 21 test sites showed that radiation levels from PAVE PAWS nowhere exceeded a fraction of a microwatt of average power. Lieutenant Colonel David Kanter, project officer for the environmental impact statement, then assured the audience that "no reliable evidence" existed to show that radiation from PAVE PAWS would have any ill effects on the surrounding population.

As the hearing progressed, a dozen or so speakers from the audience criticized the draft impact statement. One of them pointed out that the statement highlighted biological studies of "doubtful validity" whenever they tended to support the Air Force's contention that the radar would be safe. Another noted that although none of the civilian monitors who accompanied Mitchell and the Air Force measurement team had a medical degree, or any other qualifications enabling them to pass judgment on the safety of PAVE PAWS, statements issued by them had given "ordinary people the impression that a highly qualified authority had reached the conclusion that the emissions from PAVE PAWS could do no harm." Someone else cited 1977 congressional testimony by Mitchell that the biological consequences of such radiation "may not be known for several years." This writer challenged Kanter's claim that the biological effects produced by low-level microwaves had no demonstrated medical significance, saying that these effects had included changes in the immune system and in brain chemistry. He also questioned Peter Polson, senior biomedical engineer for SRI International, about the squirrel monkey study. The exchange went as follows:

"Dr. Polson, did you make autopsies on the squirrel monkeys that died—the four out of five?"

"No," Polson replied. "We made autopsies on several of the squirrel monkeys, but not all of them."

"You mean to tell me, doctor, that you had four out of five squirrel monkeys die for unexplained reasons in the irradiated group and you didn't make autopsies of them?"

"That is correct, but let me clarify that point," Polson said. "We did not expect any deaths of these animals as a result of the microwave irradiation. It is the practice when these animals do die to put some of the carcasses into deep freeze so that autopsies can be performed. Unfortunately, several of the early deaths occurred during the weekend, and when these deaths were discovered—that is true."

Toward the end of the evening, someone from the audience asked the Air Force representatives to address the medical questions that had been raised. "I would like to see a medical doctor stand up," he said.

At that point, a Colonel Mohr, who had not yet spoken or been introduced, got to his feet. "Perhaps I should identify myself," he said. "I guess I'm the only one you haven't met. I'm a flight surgeon. I was trained nearby here at Harvard Medical School. I've been in the Air Force for twenty-two years. Currently I'm in Research and Development. I practiced medicine for about five of my twenty-two years. I'm not an expert on radiation, but I am a scientist, and I'd be pleased to answer specific questions if I can."

The speaker from the audience asked Colonel Mohr if he could "guarantee us as a medical doctor representing the Air Force of the United States that the PAVE PAWS radar facility is safe medically to human beings?"

"I very carefully did read this entire assessment," Colonel Mohr began, and went on to give a rambling reply that trailed off into a hypothetical situation in which he imagined himself sitting down with members of the audience, asking if they had ever had a headache or felt tired, and then relating these subjective complaints to microwave exposure.

"You're not answering the question," the speaker told him.

"No, I'm just giving you a preamble," Colonel Mohr said. "Now, with regard to that, I find as a physician no concrete evidence that there are irreversible hazards to health occurring at the levels that have been measured or predicted for this system. That does not say that I am one hundred percent correct, that I'm God. I do not know the future any more than you do. But the evidence to date, as I see it, at these levels do [sic] not indicate a hazard to your health."

"Am I correct in assuming that your reply is basically that you do not know?" the speaker persisted.

"I do not know," Mohr acknowledged.

# THE BEST POSSIBLE FACE

THE FINAL ENVIRONMENTAL IMPACT STATEMENT released by the Air Force in May 1979 contained more than two hundred pages of letters and comments from federal, state, and local agencies, and also from interested individuals. The tone of these letters and comments ranged from total approval of to complete disbelief in the Air Force's conclusion that PAVE PAWS posed no threat to the health of people in the surrounding communities. The three-member PAVE PAWS subcommittee of the Cape Cod Planning and Economic Development Commission wrote a letter of fawning approval, telling the Air Force that the Commission "wishes to express its appreciation for the efforts undertaken by the Air Force, its satisfaction with the testing it has monitored, and acknowledges the extremely low levels of microwave radiation reported from the field tests." However, the Association for the Preservation of Cape Cod said that the biological effects of chronic exposure to low-level radio-frequency radiation were not understood, and that the conclusion of the draft statement "that exposure to PAVE PAWS RFR (radio-frequency radiation) poses no safety risk is in reality an assumption which is unsupportable at this time." Only one Cape Cod physician—Dr. Charles D.

Johnson, of Falmouth—submitted a comment for the record. "Biological studies on mice, chicks, and monkeys which indicate that microwaves have real and serious effects on the blood-brain barrier, brain chemistry and on white blood cells are discounted in the draft EIS," Dr. Johnson told the Air Force. "The final EIS should acknowledge the importance of these studies, and admit that the long-term bio-effects of low-level RFRs are not known. It should also state that very little is known about the bio-effects of pulsed RFR of the type emitted by the PAVE PAWS radar system."

But the final impact statement did no such thing. On page 2 of the summary section, its authors declared that "no scientific evidence was discovered to indicate that any ill effects will result from long-term exposure to the PAVE PAWS emissions." They went on to say that the "relatively few retrospective epidemiology studies of health effects from RFR exposure done in the United States and the USSR are not considered evidence that PAVE PAWS emissions will constitute a hazard to the population." In this way they glossed over the results of a retrospective epidemiological study of the health experience of 1,827 State Department employees who had lived and/or worked at the American Embassy in Moscow between 1953 and 1976. That study had been conducted by Dr. Abraham Lilienfeld and some colleagues at the Johns Hopkins University School of Public Health, who found that female employees of the State Department had experienced an unusually high death rate from cancer. The authors of the impact statement sloughed over Lilienfeld's finding by claiming that, "with the exception of cancer-related deaths among female employee groups (both Moscow and control), mortality rates for both Moscow and control groups were less than for the U.S. population at large." This, however, is how Lilienfeld had described his finding: "A relatively high proportion of cancer deaths in both female employees groups was noted—8 out of 11 deaths among the Moscow [group] and 14 out of 31 deaths among the Comparison group." Lilienfeld added that it was not possible "to find any satisfactory explanation for this, due mainly to the small number of deaths involved and the absence of information on many epidemiological characteristics that influence the occurrence of various types of malignant neoplasms."

Lilienfeld's study turned out to be seriously flawed, because his control group consisted of State Department employees who had worked at embassies and consulates in other Eastern European countries, where microwave surveillance operations had also been undertaken. These employees should not, of course, have been considered as an unexposed comparison population. Moreover, Lilienfeld had underestimated the health hazard of microwave radiation at the embassy by failing to point out that the death rate—particularly from breast cancer—was extraordinarily high among women living in apartments on the third through the seventh floors of the embassy during the late 1960s and early 1970s, when the irradiation of the embassy had become a source of major concern to the State Department and the CIA. These women, it has been suggested, had been exposed not only to low-level radiation beamed into the building by the Soviets, but to much higher level radiation emitted by electronic countermeasures equipment that had been used by American intelligence agencies in an effort to jam the Soviet transmitters.

After declaring that there was no epidemiological evidence that human beings had ever suffered from exposure to radiofrequency radiation, the authors of the Air Force's environmental impact statement proceeded to buttress their assertion by announcing that the National Academy of Sciences had concluded that "it is improbable that exposure will present any hazard to the public." The Academy had already submitted its own report on PAVE PAWS to the Air Force, and the Air Force had plucked this phrase from its concluding section in such a way as to make it not only self-serving but also misleading. The Academy's report was entitled "Analysis of the Exposure Levels and Potential Biologic Effects of the PAVE PAWS Radar System." The final paragraph of chapter 3, "Summary and Conclusions," which appears on page 81, reads as follows:

> In conclusion, the PAVE PAWS radar may be anticipated to expose a limited number of members of the general public intermittently to low intensities of pulse-modulated microwave fields with maximal instantaneous intensities of 100 uW/cm² [100 microwatts per square centimeter] or less and time-averaged intensities lower by two orders of magnitude. There are no known irreversible effects of such exposure on either morbidity or mortality in humans or other species. Thus, it is improbable that exposure will present any hazard to the public. In view of the known sensitivity of the mammalian CNS [central nervous system] to electromagnetic fields, especially those modulated at brainwave frequencies, the possibility

cannot be ruled out that exposure to PAVE PAWS radiation may have some effects on exposed people. Because these effects are still hypothetical, it is not feasible to assess their health implications. Such assessment will require additional research and surveillance and must be addressed in future evaluations of the potential exposure effects of PAVE PAWS and other high-power-output radar systems.

Thus did the National Academy of Sciences acknowledge that members of the general public exposed to radiation from PAVE PAWS would become participants in a biological experiment whose outcome would be ascertained by future evaluation. Some of the rest of chapter 3 was equally unsettling in its assessment of the future well-being of Cape Codders who would be living in the spillover of the giant radar's beam. On page 80, for example, the authors of the report got down to the specifics of the brain-wave frequency that had been used by Adey and Bawin in their pioneering experiments. "Data from both *in vivo* and *in vitro* studies suggest maximal sensitivity of neural systems to fields modulated at mammalian brain wave frequencies (i.e., 1–20 Hz), which include the predominant PAVE PAWS modulation frequency at 18.5 Hz," they wrote. "Because of the aforementioned possibility of field-intensity windows and the lack of adequate data on mammalian systems, it is not known whether such effects will be induced in humans under the anticipated exposure conditions."

The report went on to put the best possible face on this scary possibility. "The effects of such exposures of members of the public, if they occur, will, on the basis of available data and the known interaction mechanisms with biologic systems, be reversible or transient," its authors wrote reassuringly. "Thus, the possible exposure effects of PAVE PAWS should be restricted to transient, reversible functional alterations in the CNS that may or may not be perceived by the exposed persons." Having suggested a new version ("what you don't perceive won't hurt you") of an old folk saying, the writers paused in the middle of their headlong plunge down the path of wishful thinking and, as if worried by the absurdity of their conclusion, qualified it with a confession of ignorance. "Whatever the effects of exposure on the human central nervous system are, it is not known whether the effects are deleterious to health," they declared. "It has not been established, for example, that such effects involve impairment of judgment or alterations in mood that would impose psychologic or physiologic burdens on those affected."

Although copies of the Academy's report were available in some Cape Cod libraries, it is not known how many local people took the time and trouble to read it, or how many of them ever heard about the admission of the committee members that they had no idea what the ultimate effects of the radiation emitted by PAVE PAWS upon human health would be. It is a fact, however, that no newspaper on the peninsula saw fit to report on, let alone reprint, this admission.

During the summer and autumn of 1979, the will and the wherewithal to resist the imminent startup of the radar dissipated like smoke in the wind. In September, a group called Sandwich Citizens Rallied Against PAVE PAWS revived the lawsuit that had been filed by the Cape Cod Environmental Coalition, and asked the Federal District Court to find that the Air Force's final environmental impact statement was inadequate. In early December, however, Federal District Judge Joseph Tauro granted a motion by the Air Force to dismiss the case. In a four-page decision Judge Tauro ruled that "the relevant provisions of the National Environmental Policy Act do not empower this court to make substantive decisions regarding technical matters disputed by the parties." He also ruled that two affidavits stating that microwave radiation might be a health hazard were not relevant, saying that they represented "the speculation of experts." (One of the affidavits had been given by Ruth Hubbard, a professor of biology at Harvard.)

It is not known whether Judge Tauro read the conclusions of the National Academy of Sciences report before reaching his decision, or, if he did, whether he found them to be irrelevant as well. It is known, however, that the PAVE PAWS radar system had already become operational, and had begun to irradiate the inhabitants of the four towns that surround it—Sandwich, Mashpee, Bourne, and Falmouth—with low-level microwave radiation that was being pulsed at a frequency and modulation that had been shown in repeated laboratory experiments to be capable of changing the chemistry of brain cells in living creatures.

# A GRAVE LACK OF
# GOOD JUDGMENT

BACK IN EARLY FEBRUARY 1979, Adey had written to John M. Richardson, the chief scientist of the National Telecommunications and Information Administration and the chairman of its Electromagnetic Radiation Management Advisory Council, and sent copies to Carlos Stern, an Air Force environmental official, and several other members of the Council. Adey told Richardson that each PAVE PAWS transmitter operated in excess of 1,000 megawatts of time-averaged power, and that its sidelobes—the sidewise radiation spilling from the main beam, which could be expected to irradiate inhabited areas—"require careful evaluation from the viewpoint of human safety, particularly in the light of a system life-expectancy of up to 20 years." He advised Richardson that the modulation characteristics of any radio-frequency signal "are powerful determinants of the nature and extent of tissue interactions," and that "this aspect of PAVE PAWS requires careful evaluation since certain of its modulation modes closely resemble those known to react strongly with tissue of the central nervous system." He added, "The Air Force's need for this type of radar is unquestioned, but the choice of the Cape Cod site may reflect a grave lack of good judgment."

Adey apparently hoped that by sending Stern a copy of his letter to Richardson he could influence the Air Force and SRI to do a more thorough and evenhanded job of rewriting the draft impact statement. However, the authors of the final statement simply added brief references to studies they had omitted—among them a single-sentence reference to Adey's cat study—in the middle of a section entitled "PAVE PAWS and Safety to Human Populations." They concluded, "We see no evidence that the levels of general public exposure to PAVE PAWS RFR are hazardous."

As for the National Academy report, Adey and his fellow committee members apparently hoped that by expressing uncertainty about the ultimate effect of PAVE PAWS radiation on human health they could influence the Air Force and SRI to acknowledge as much in the final impact statement. This did not happen. Instead, their carefully qualified doubts merely served as camouflage for the fact that, when all was said and done, they were conferring the imprimatur of the National Academy of Sciences upon the installation and operation of the radar system.

In early January 1980, Adey appeared at the annual meeting of the American Association for the Advancement of Science in San Francisco as an invited speaker at a symposium on the biological effects of microwave radiation. At that time, he described some recent studies he had conducted with Richard A. Luben, a cell biologist at the University of California, Riverside, which demonstrated that low-frequency magnetic fields pulsed at either fifteen or seventy-two times a second could affect the ability of parathyroid hormones to trigger the activity of adenylate cyclase, a metabolic enzyme inside the cell membrane which plays a crucial role in the formation of new bone. Adey said that the power of the external fields was far too low to alter cell chemistry by itself, but he suggested that the ability of the cell membranes to sense the weak external signals might trigger large changes in cell behavior. He emphasized that he and his associates had observed the changes in calcium binding and hormonal activity only when the external fields were generated at specific power levels and were modulated at certain brain-wave frequencies.

Toward the end of his presentation, Adey unleashed a heavy barrage of criticism at the Air Force, saying that it had shown "a lack of integrity" in its assessment of the possible health hazards of the radiation emanating from PAVE PAWS. He pointed out that the characteristics of the electromagnetic radiation he had used in his experiment were almost identical to those of the radiation being beamed from the radar; and he went on to suggest that the Air Force had tried to discount the value of his work by claiming that it had been performed with isolated chick-brain cells, although Air Force officials could easily have learned that he and his colleagues had produced the same chemical changes in the brains of living cats.

Following the San Francisco meeting, articles about Adey's pioneering research and his denunciation of the Air Force appeared in a number of newspapers across the nation—among them the San Francisco *Chronicle* and the Boston *Globe.* Curiously, not a word about it found its way into the Cape Cod *Times* —the Cape's largest newspaper, with a circulation of 37,000. Indeed, the Cape Cod *Times* had carried very little news about the radar since running an editorial on June 2, 1979, which expressed the hope that no major flaws would be found in the Air Force's final environmental impact statement, and that "all can agree PAVE PAWS poses no major health or safety threat to the community."

Almost no mention of the health effects of PAVE PAWS appeared in the Cape Cod *Times* during 1980 and 1981. On December 31, 1982, however, the newspaper ran another story about the radar. It appeared under a headline that read "PAVE PAWS: WHERE HAS THE CONTROVERSY GONE?" and contained mention for the very first time of the reservations about the safety of the radar that had been expressed three and a half years earlier by the authors of the National Academy of Sciences report. The story quoted John Mitchell, who pointed out that the radiation levels from PAVE PAWS measured by the Air Force were extremely low. Once again, Mitchell refused to guarantee that these levels posed no health hazard. "There's no way you can guarantee anything," he declared. He then told the Cape Cod *Times* that scientists at the University of Washington in Seattle were conducting a study of the long-term effects of low-level microwave radiation. This study was being financed by the Air Force to the tune of some $4 million, and was being carried out by Professor Arthur W. Guy, an engineer who is director of the Bioelectromagnetics Research Laboratory at the University of Washington's School of Medicine. Chung-Kwang Chou, associate director of the laboratory, told the Cape Cod *Times* that the Seattle study consisted of exposing a hundred rats around the clock for two years to low levels of microwave radiation similar to those being emitted by PAVE PAWS. Chou said that the exposed rats would be examined and compared to a non-irradiated control group, in order to determine whether the radiation had affected their blood chemistry, body weight, and behavior.

Meanwhile, Adey was pushing on with research aimed at finding out how weak electromagnetic fields might trigger amplified changes in brain chemistry. Since his earlier work with Leonard Kaczmarek on calcium ions had suggested that highly non-linear effects must be involved—effects that would allow large releases of chemical energy to be triggered by tiny stimuli—he now proposed a three-stage model based upon the amplification of weak cell-surface events. In the first stage, the cell membrane senses the arrival of a weak field and reacts to it by initiating, in domino fashion, a release of calcium bound to the protein strands that protrude from the membrane surface. (Because of the domino effect, the energy released by the altered calcium binding is far greater than that of the initiating chemical or electric triggers.) In the second stage, this released energy is transmitted in the form of signals that travel along the protein strands to the interior of the cell. In the third stage, the cell interior reacts to the arriving signal by stimulating or inhibiting intracellular enzymes—protein molecules that act as chemical catalysts.

During 1981 and 1982, Adey and his colleagues conducted a series of experiments which furnished further evidence to support the theory that electromagnetic fields could activate calcium-dependent enzymes in cells, and cast additional doubt upon the safety of PAVE PAWS. Indeed, while the Cape Cod *Times* was asking where the controversy over PAVE PAWS had gone, Adey and Daniel B. Lyle, a research immunologist at the University of California, Riverside, together with two associates, were demonstrating that the PAVE PAWS carrier frequency of 450 MHz modulated at the electrical distribution system frequency of 60 hertz could significantly suppress the ability of cultured T-lymphocyte cells from mice to kill cultured cancer cells. When the four researchers published their findings in the journal *Bioelectromagnetics* in 1983, they noted that progressively smaller suppression occurred when the microwave radiation was modulated at 40, 16, and 3 hertz. There were, of course, possibly serious implications in this—not only for the tens of thousands of people living close to PAVE PAWS but also for the millions of people living in the vicinity of the power lines and electrical distribution wires. In fact, when Lyle later repeated the study, using simulated 60-hertz high-voltage power-line fields, he again observed a large decrease in the killing capacity of T-lymphocyte cells—a chilling finding when placed in the context of Wertheimer and Leeper's pioneering study of the increased incidence of cancer in children who had been exposed to low-level magnetic fields from high-current electrical distribution wires.

The results of a study performed in 1983 by Craig V. Byus, an associate professor of biochemistry in the Department of Biochemistry at the University of California at Riverside, together with Adey and two researchers from Byus's department, gave cause for further concern about the future well-being of the inhabitants of Upper Cape Cod. Byus and his colleagues exposed cultures of human tonsil lymphocytes to low-level 450-MHz radiation modulated at ELF frequencies ranging from 3 to 100 hertz. They found that the activity of certain protein kinase enzymes within the lymphocyte cells was sharply reduced when the microwave field was modulated at frequencies of 16 to 60 hertz. Because the effect upon the activity of the protein kinase was transient, Byus and Adey were cautious in their assessment of it when they and their associates published these findings in *Bioelectromagnetics,* in 1984. "At present, our data offer no insight into the real biological effect of these fields upon lymphatic functions in particular, or upon the general state of the immune system," they wrote. "We cannot say at this time whether this transient decrease in protein kinase alters the ability of lymphocytes to perform any special function. For this reason, it is difficult to make any statement concerning potential damage to the cell based upon these data."

Byus and Adey pointed out that recent studies which concluded that existing microwave standards in the United States were safe had ignored the effects of ELF modulation. They then warned that their findings with human tonsil lymphocytes, together with the findings of the earlier mouse lymphocyte study conducted by Adey and Lyle, "indicate the need for a more cautious assessment, with due consideration of amplitude modulation found in virtually all domestic, industrial, and military microwave fields."

How the unsuspecting residents of the four towns surrounding PAVE PAWS, who were then entering their fifth year of exposure to low-level, ELF-modulated 450 MHz radiation, might have re sponded to all of this must remain a matter for conjecture, because no one bothered to inform them about it. For that matter, the editors of the Cape Cod *Times* did not even bother to follow up on the story the newspaper had published on December 31, 1982, about the Air Force–financed study under way at the University of Washington to determine the long-term effect upon laboratory rats of radiation similar to that emitted by PAVE PAWS. Had they done so, they could have learned of a highly revealing panel discussion that took place in Boulder, Colorado, on June 16, 1983, at the annual meeting of the Bioelectromagnetic Society—a multidisciplinary organization that had been formed in 1979 to study the biological effects of environmental electromagnetic fields. At that time, Chung-Kwang Chou reported some preliminary results of the study that he and Guy had been conducting. According to Chou, it had been determined midway through the experiment that rats exposed to ELF-modulated microwave radiation had increased numbers of B- and T-type lymphocytes, as well as enlarged adrenal glands.

On August 18, 1984, the Cape Cod *Times* published a front-page piece about the final results of Guy and Chou's study. It started out by saying that "A five-year study on the long-term effects of low levels of microwave radiation has sparked a new debate over the safety of the Air Force's PAVE PAWS microwave radar station in Bourne." The article went on to quote Ruth Hubbard, the Harvard biology professor whose affidavit warning of the dangers of PAVE PAWS radiation had been dismissed as irrelevant five years earlier by Federal District Judge Tauro. Professor Hubbard suggested that people living near the giant radar had become guinea pigs, and she called upon the Air Force to conduct a study of the long-term biological effects of PAVE PAWS radiation on those people. The *Times* then quoted the ever optimistic John Mitchell, of the Air Force's School of Aerospace Medicine, who declared, "There's no question that PAVE PAWS is safe," and that the radiation levels from PAVE PAWS were so low that such a study "would not be practical." Once again, however, Mitchell refused to say that there would be no long-term adverse health effects caused by microwave radiation from PAVE PAWS. "You can't guarantee anything today," he told the newspaper. Tucked away at the end of the article was the revelation that sixteen malignant tumors had been observed among rats that had been exposed to radiation similar to that emitted by PAVE PAWS, as compared with only four cancers among non-irradiated control animals.

By contrast, a story about the same study that had appeared a day earlier on the front page of the Boston *Globe* went into detail about a report in the newsletter *Microwave News,* which disclosed for the first time that no fewer than seven of the tumors in the exposed animals involved the endocrine system, as compared with only one endocrine tumor in the controls, and that seven of the exposed animals had developed benign adrenal tumors, while one had been found in the control rats. On the same day that the *Globe* published this startling information, an Associated Press report carrying similar information about the study appeared in newspapers around the nation and the world. It quoted Dr. Samuel Milham, Jr., of the Washington State Department of Social and Health Services, who had recently found increased rates of leukemia among power station operators, telephone linemen, and other workers chronically exposed to electric and magnetic fields. "It looks like the microwave radiation may have been a tumor promoter," Milham said. "It didn't initiate the cancers, but once some cancer cells got going it promoted growth of the tumors."

In light of the worldwide attention accorded the University of Washington study, one can only wonder what the editors of the Cape Cod *Times*—the one newspaper in the nation whose readers should have been informed in minute detail about the findings of Chou and Guy—were thinking of when they omitted so many important aspects of the story. Whatever it was, it seems unlikely that they did not know that virtually every leading newspaper in the nation—and the *CBS Evening News* as well—had covered the story and its profoundly serious implications for the public health in far more depth than they had.

As for the real story behind the announcement of the results of the University of Washington study, it emerged in a 1,500-word account by Louis Slesin that appeared in the July–August 1984 issue of *Microwave News.* Slesin, who was present at the meeting where the study results were announced, revealed that Chou, Guy, and Lawrence Kunz, a veterinary pathologist at the university, were trying to cast doubt upon their findings and to claim that their study had produced negative results. In the end, the three researchers failed to publish the study in a peer-reviewed medical journal. Instead, the Air Force buried the $4 million study in a 4,000-page, nine-volume tome that had to be specially ordered from its School of Aerospace Medicine. Moreover, in spite of the fact that Guy and his colleagues subsequently found two more primary tumors among the exposed rats—thus raising the total to eighteen tumors in the exposed animals, as compared with only five among the controls—he and his co-workers found themselves able to conclude in the summary chapter of the ninth volume that they could identify "no defendable trends" in the development of cancer among the exposed animals. Except for one mention in passing, however, not another word about this study ever appeared in the Cape Cod *Times*.

Six months later, the findings of Chou and Guy's rat cancer study were supported by those of the largest epidemiological investigation of the effects of radio-frequency and microwave radiation ever undertaken. This was a two-year survey of cancer in Polish military personnel conducted by Dr. Stanislaw Szmigielski, an internationally known researcher on the biological effects of non-ionizing radiation, and some colleagues at the Center for Radiobiology and Radioprotection, in Warsaw. Szmigielski and his associates compiled all the cases of cancer that had been reported in the Polish military services between 1971 and 1980, and then analyzed them in terms of the exposure of each individual to radio-frequency and microwave radiation from radar and other sources. They found that servicemen exposed to non-ionizing radiation were almost seven times as likely to develop cancer of the blood-forming organs and lymphatic tissue as those who were not exposed, and that the odds of their developing thyroid tumors was more than four times as great. They found that younger exposed personnel—those between the ages of twenty and twenty-nine—had a 550 percent greater risk of being stricken with cancer than their unexposed counterparts. Over all, servicemen who worked with or near radiation-emitting devices were more than three times as likely to develop cancer as unexposed personnel in other military occupations.

Slesin, to whom Szmigielski had sent the results of his yet unpublished study, believed them to be so important that he published and distributed a news release about them to newspapers and magazines across the nation. As a result, a story appeared on the front page of the Boston *Globe* on March 15, 1985, beneath the headline "STUDY LINKS CANCER, MICROWAVE RADIATION." Boston is only 65 miles from Hyannis, where the Cape Cod *Times* is written and published, and it is safe to say that every senior editor on that newspaper reads the *Globe* every day, to keep up with important national, international, and regional events. Not a word about Szmigielski's findings ever found its way into the Cape Cod *Times*, however, except for a passing reference in June 1986, when the newspaper mentioned a Polish study as being one of several that "link long-term exposure to low levels of microwaves to certain health disorders, including cancer, high blood pressure, headaches, loss of memory, and brain damage."

---

# UNDER THE LIGHT BULB

BY 1986, THE SUBJECT OF CANCER was very much on the minds of the residents of the Upper Cape towns of Sandwich, Bourne, Falmouth, and Mashpee. In March of that year, the Massachusetts Department of Public Health released a report revealing that between 1979 and 1981, women living in the four towns had died of leukemia at a rate that was 23 percent higher than that of other women in Massachusetts, and had died of cancer of the liver, bladder, and kidney at a rate that was 61 percent higher. An accompanying memorandum showed that between 1982 and 1983, women living in the four towns had developed liver, bladder, and kidney cancers at a rate almost four times as high as that of women living in the eleven other towns on Cape Cod, and that during this period men and women living on the Upper Cape developed cancer of all types at a rate sixteen percent higher than that of the residents of the rest of the Cape.

It soon came to light that the Air Force and the Massachusetts Air National Guard had been dumping toxic chemicals into waste-disposal sites on the base for 46 years. The fact that some of these chemicals were already contaminating groundwater supplies became evident when volatile compounds, such as chloroform,

trichloroethylene, and tetrachloroethylene—the last two, suspected carcinogens both, are solvents widely used to degrease
airplane engines and other machinery—turned up in a number of
private wells in the Ashumet Valley, close to the southeastern
corner of the base. As a result, several hundred residents of the
Valley were advised not to drink their water or use it for cooking.

Late in March, some Cape residents who were attending a
public meeting about the Upper Cape's elevated cancer rate
brought up the subject of whether radiation from PAVE PAWS
could have anything to do with it. They were told by state officials
that environmentally induced cancer takes at least ten years to
develop. The officials implied that the radar facility had not been
in operation long enough for this to occur. (They were obviously
unaware that if radiation from PAVE PAWS were acting not as a
tumor initiator, but as what Milham called a tumor promoter, the
latency period for the development of cancer could be less than
ten years.)

In June 1986, however, the question was raised again—this
time in an article in the Cape Cod *Times* under the headline
"PAVE PAWS BEAMS STILL SPARK FEAR IN NEIGHBORS." The article quoted a variety of people who were either for or against the
radar. Susan Klein, director of the Sturgis Library in Barnstable,
who had opposed the installation of PAVE PAWS in the late
1970s, pointed out that the Air Force had never answered the
health questions about the radar that had been raised eight years
earlier. Ruth Hubbard, the Harvard biology professor who had
also opposed the radar, repeated her suggestion that Upper Cape
residents were "guinea pigs," because the long-term effects of
microwave radiation were largely unknown. Jerome Krupp, chief
of biological effects for the Radiation Sciences Division of the Air
Force School of Aerospace Medicine, tried to play down the
results of the study conducted for the Air Force by Guy and
Chou, saying that the rats had been subjected to constant microwave exposure under a "worst-case situation," and that "there's
no one living under those conditions." (He chose to ignore the
fact that because PAVE PAWS is constantly emitting low-level
microwave radiation, the inhabitants of the Upper Cape are constantly being exposed to it.) According to Krupp, the microwave
levels emanating from PAVE PAWS were far too low to be a
health hazard. "It's almost like sitting under a light bulb," he
said.

During the summer of 1986, it became apparent that chemical contamination at the Otis Air National Guard Base and Camp Edwards was far worse than anyone had suspected. On June 30, the National Guard Bureau, the federal organization of the National Guards, held a press conference to announce that the E. C. Jordan Company of Portland, Maine—an environmental consulting firm it had hired to evaluate the military reservation—had found no fewer than forty-six significant hazardous waste sites there. According to E. C. Jordan, monitor wells installed at the base in 1985 had detected a level of tetrachloroethylene (a liver cancer–producing agent) almost five times greater than the safety limit recommended by the federal government. The company also said that more than a million gallons of aviation fuel had been dumped into the ground during the maintenance and testing of a squadron of EC-121 early warning radar aircraft stationed at Otis from 1955 until 1970.

In January 1987, the Massachusetts Department of Environmental Quality and Engineering announced that it was classifying the entire 21,000-acre military reservation as a hazardous toxic waste disposal site. According to the department, as much as 1,337,000 gallons of fuel had been spilled into the ground at the base over the years, and there had been eighteen known spills of toxic chemicals. Department officials also revealed that military personnel operating an inspection laboratory at the base between 1955 and 1970 had used a dry well to dispose of nearly 7,000 gallons of trichloroethylene cleaning solvent.

Meanwhile, back in September, the Air Force, obviously worried that microwave radiation from PAVE PAWS might be implicated in the high cancer rates among Upper Cape residents, departed from its long-standing position of refusing to monitor radiation levels, and measured them at fifteen different places in the surrounding towns. On September 22, 1986, Lieutenant Colonel Gayle White, the commander of PAVE PAWS, said that the data so far showed microwave levels to be only one thousandth the recommended state and federal safety standards, which is similar to what John Mitchell had told Cape Codders eight and a half years earlier. White said that the measurements should help to prove that the radar was not contributing to the unusual increase of cancer among residents of the Upper Cape. "They can put that issue to bed," he declared. A year later, in October 1987, the Department of Public Health announced that the Upper

Cape's unusually high cancer incidence had continued during 1985, and that the residents of Barnstable, to the east of the radar, as well as Bourne, Falmouth, Mashpee, and Sandwich had a seventeen percent higher rate of cancer than other residents of Massachusetts.

In the spring of 1987, it came to light that the Air Force had been planning to quadruple the power output of the Cape Cod PAVE PAWS but had decided to hold off for a while. At about the same time, the Air Force announced that the maximum time-averaged radiation level found in the communities surrounding PAVE PAWS was slightly less than one and a half microwatts per square centimeter. The Air Force said that none of the other microwave power densities measured in the four towns exceeded three tenths of a microwatt per square centimeter at a height of six feet.

All this sounded fine, but in fact time-averaging had skewed the latest Air Force measurements of the radiation from PAVE PAWS in the same way that it had skewed the measurements in the summer of 1978. Instead of calculating the non-thermal effect of high-energy peak pulses of radiation which have been shown to be capable of penetrating deeply into tissue, altering the chemistry of the brain, and disrupting the immune system, the Air Force again chose to assess the effects of the radiation on human health only in terms of the thirty-four-year-old thermal theory, which was now considered inappropriate by many of the leading scientists in the field. In short, the Air Force was citing an obsolete exposure guideline in order to assure the residents of Upper Cape Cod, who surely knew it to begin with, that they were not being cooked. Skeptics who had seen the Air Force pull this stunt time and again over the years called it the English muffin standard, because of the implication that microwave radiation was harmless as long as you didn't turn brown or feel toasty.

As for Jerome Krupp's contention that exposure to microwave levels emanating from PAVE PAWS was no more harmful than sitting under a light bulb, it was called into question in the summer of 1988, when it was revealed that the Air Force was considering relocating a PAVE PAWS radar it had constructed at Robins Air Force Base in Georgia, in 1986, in order to reduce the danger of explosion in airplanes flying through its main beam. According to Raytheon, the manufacturer of PAVE PAWS, the high energy contained in the radar's peak pulses "may pose a hazard to electro-explosive devices (EEDs) carried on military and commercial aircraft." As a result Raytheon warned that airplanes carrying such devices should stay at least one nautical mile from the radar, and that when the Air Force increased the power output of PAVE PAWS, the air-space "hazard zone" should be increased to three nautical miles. The company further proposed to install a special tracking device within each of the radar's two arrays of radiating elements, which would monitor air traffic approaching the hazard zone and order an automatic shutdown of PAVE PAWS by reducing the power of its pulses. Raytheon said nothing about how the high-energy peak pulses of PAVE PAWS might be affecting the delicate electrochemical balance of the human brain.

# A QUESTION OF PROMOTION

THANKS TO GOVERNMENTAL FOOT-DRAGGING, the reason why people living in towns adjacent to the PAVE PAWS radar are developing cancer at a rate far higher than other people living on Cape Cod and in Massachusetts may not be found for some time to come. Moreover, because of the Air Force's policy of dumping millions of gallons of aviation fuel and other toxic waste into the sandy aquifer, thus contaminating groundwater on Upper Cape Cod, scientists trying to solve the mystery will have to take many factors into consideration. One such factor—in spite of the National Academy's premature attempt to exonerate it, and the Air Force's ludicrous effort to equate its effects with those of the incandescent light bulb—will be the question of whether chronic exposure to low-level radiation from PAVE PAWS has acted to promote the incidence of cancer in people who are already at risk because they have been exposed to cancer-producing chemicals.

Reason for suspecting that this may be the case can be found in the 1983 discovery by Adey and Byus that 450 MHz radiation modulated at the ELF frequency of 16 hertz—radiation similar to that being generated by PAVE PAWS—sharply reduces the activity of certain protein kinase enzymes within human lymphocytes. This finding raises disturbing questions about the effect of such radiation upon the human immune system, which is the body's first line of defense against the development of cancer. The questions arise as a result of the work of Yuri Nishizuka, a cancer researcher at the University of Tokyo, who demonstrated in 1983 that one of the affected protein kinase enzymes, known as protein kinase C, is a specific receptor at the cell membrane for a class of extremely powerful cancer-promoting substances called phorbol esters.

Phorbol esters are a component of croton oil, which is found in the seeds of *Croton tiglium L.*—a vegetable plant in the Euphorbiaceae family. The oil was discovered to be a cancer-promoting agent in 1941, when a researcher observed that if he applied it, together with a known carcinogen, to the backs of mice, more tumors occurred than with the carcinogen alone. In 1944, other researchers discovered that an application of croton oil caused cancers to develop in mice that had already been exposed to a subcarcinogenic dose of a cancer-producing chemical. During the early 1980s, Japanese researchers found that the inhabitants of several regions in southeastern China who are exposed to croton oil because they cook and eat large quantities of vegetable plants in the Euphorbiaceae family, were developing cancer of the nasal passages at a very high rate.

Adey points out that it is important to distinguish between substances that promote cancer and those that initiate it. "The accepted model of tumor formation involves at least two and possibly three stages," he said recently. "They are initiation, promotion, and progression. Initiation occurs when the DNA in the nucleus of a cell is damaged by ionizing radiation, such as gamma rays and X-rays, or by certain chemicals—mostly coal-tar derivatives—such as benzo-a-pyrene, which is found in cigarette smoke. The damaged cell has undergone mutation, and is now capable of passing its damaged DNA to daughter cells. It is therefore a cancer cell, but it may not necessarily form a tumor. The next phase in carcinogenesis is promotion with tumor formation. This involves unregulated growth. Cancer promoters include plant derivatives, such as phorbol esters, tobacco proteins, polychlorinated biphenyls, called PCBs, DDT, and other chlorinated hydrocarbons."

A leading theory for cancer promotion has been advanced by James E. Trosko, a geneticist in the College of Human Medicine at Michigan State University in East Lansing; by Andrew Murray, a cell biologist at the University of Adelaide, in Australia; and by Hiroshi Yamasaki, a cell biologist at the International Agency for Research on Cancer, in Lyon, France. They hypothesize that cancer promotion often involves a failure in cell-to-cell communication. It has long been known that ordinary, healthy cells grow rapidly until they begin to touch. At that point, tiny protein particles appear between them, which enable the cells to communicate electrically and chemically. These small protein particles are perforated by many tiny tubes or canals that allow essential substances made in one cell to pass to another, and they form a system that is called gap-junctional communication—a concept that has been postulated and verified by Werner Loewenstein, an electrophysiologist at the University of Miami. If the communication provided by the protein particles is blocked by phorbol esters, or other tumor promoters, the result is unregulated cell growth. Conversely, if pre-malignant cells are brought in contact with normal cells that are still capable of maintaining gap-junctional communication, the pre-malignant cells cease to grow at the points of contact. According to Adey, this provides a dramatically new vista on the essential biology of cancer and tumor formation. "The positive side is that new forms of therapy may now be developed which could be used to increase gap-junctional communication in malignant cells, with the expectation that unregulated growth would cease and that the malignant cells would become healthy again," he pointed out recently. "Such a possibility contrasts starkly, of course, with current concepts of chemotherapy, which are based upon the total destruction of all cancer cells."

When Adey and Byus learned of Nishizuka's finding that phorbol ester specifically modifies the activity of protein kinase C, especially in combination with ELF-modulated fields, they recognized its significance for their further studies. "As it happened, Craig Byus had had long experience in assaying an enzyme called ornithine decarboxylase—ODC for short—which is essential for growth in all cells," Adey has explained. "ODC participates in the synthesis of DNA protein by forming polyamines, which are the building blocks for DNA. For some time now, it has been known that all cancer promoters increase ODC activity. However, not all substances that increase ODC activity are cancer promoters. Nevertheless, in clinical studies high levels of ODC are a reliable index of malignancy—as, for example, in the human prostate. In 1985, Byus and I, together with Susan E. Pieper and Karen Kartun, both of whom were junior researchers in Byus's lab, found that when we modulated a four-hundred-and-fifty-megahertz microwave field at sixteen hertz, and generated it at a power intensity of one milliwatt per square centimeter, we could increase ODC activity up to fifty percent in cultured liver and ovary cells of Chinese hamsters and in cultured human melanoma cells. We also found that when the ELF-modulated field and the phorbol ester acted together, their joint effect on ODC activity was about twice that of either alone. As a result, we cautiously hypothesize that exposure to low-level radiation and environmental chemical pollutants may be joint factors in increasing the incidence of human cancer. It should be emphasized that our suggestion is speculative, and must be substantiated through experimental evidence that can only be obtained by conducting extensive animal studies."

The extraordinary discovery by Byus, Adey, and their two young colleagues that radiation similar in carrier frequency and modulation characteristics to that emanating from PAVE PAWS increases ODC activity in cultured human melanoma cells, and may therefore promote cancer, was published in August 1988 in *Cancer Research*. However, because of the difficulty in obtaining research funds, only one animal experiment is under way to substantiate their four-year-old finding—unless one believes that an experiment is being conducted upon the thousands of human beings residing in Sandwich, Mashpee, Bourne, and Falmouth, who have undergone chronic exposure to low-level, ELF-modulated radiation from PAVE PAWS for nearly a decade, and have been found to be developing cancer at an unusually high rate.

Even more disturbing are the findings of a subsequent study that Adey, Byus, and Pieper conducted in the winter of 1986. At that time, the three researchers discovered that a one-hour exposure to a 60-hertz electric field of between one tenth of a millivolt and 10 millivolts per centimeter produced a fivefold increase in ODC activity in cultured human lymphoma cells. They also found that the same power-line field produced a two- to threefold increase in ODC activity in mouse myeloma cells after one to two hours of exposure. These results were published in *Carcinogenesis* in October 1987.

"We have no explanation for the different responses to the

sixty-hertz field on the part of different cell types," Adey says. "We feel that it is highly significant, however, that only short, intermittent exposure to the sixty-hertz field was required to elevate ODC activity. We can now hypothesize that exposure to low-energy fields, such as those emanating from power lines, may provide a tumor-promoting stimulus, in the manner of phorbol ester compounds. Once again, it should be emphasized that such a relationship is based upon indirect evidence, and awaits direct experimental confirmation."

Adey's willingness to hypothesize that exposure to weak 60-hertz electromagnetic fields, such as those surrounding power lines, may promote cancer is something that should be taken very seriously by public health officials across the nation. After all, an alternating-current electrical field of one tenth of a millivolt is present at all times in the tissue of a human being who is standing beneath a typical overhead high-voltage power line.

# CORROBORATION

ALTHOUGH THE PIONEERING EXPERIMENTS of Adey and his colleagues have been accepted and praised by many of their peers in the medical and scientific community, there have been relatively few attempts to replicate them. A significant exception has been the work of Carl F. Blackman, a research biologist in the Developmental and Cell Toxicology Division of the EPA's Health Effects Research Laboratory at Research Triangle Park, in North Carolina. Blackman is a tall, lean man of forty-eight, with a Lincolnesque face and a serious demeanor. Born in Annapolis, he went to Colgate, where he studied physics and mathematics. Then he studied at Pennsylvania State University, where he received his M.S. and Ph.D. in biophysics. Afterward, he did postdoctoral work in the molecular biology of the aging process at the Brookhaven National Laboratory in Upton, New York. In 1970, he moved to the FDA's Bureau of Radiological Health in Rockville, Maryland, where he began to investigate the genetic effects of microwave radiation. A few months later he was transferred to the recently created EPA, where he continued this line of research.

In 1974, Blackman and several of his colleagues attended a

meeting at the New York Academy of Sciences. There, they heard Adey's associate, Suzanne Bawin, discuss the experiments she had conducted with Adey, showing that a 147 MHz ham radio carrier wave, when modulated at brain-wave frequencies, could alter the outflow of calcium ions from chick-brain tissue, and that this phenomenon occurred most readily at sixteen hertz. It was an event that profoundly altered the course of Blackman's research career.

"Up to that time, I had been studying the biological effects of microwave radiation the same way practically everybody else had —with 2,450-megahertz unmodulated radiation generated by a magnetron from a microwave oven system—and, like practically everybody else, I believed that the only major biological effect of microwaves was their ability to heat tissue," he remembers. "So you can imagine how bowled over I was by what Suzanne Bawin was saying. After all, if the biological effects she was describing had been caused by simple heating, the chick-brain tissue would not have responded differently to different modulation frequencies."

In 1976, Blackman and some colleagues at the EPA were able to confirm Bawin and Adey's finding of a 16-hertz frequency "window" in which the outflow of calcium ions from chick-brain tissue was most likely to occur. At the same time, they expanded upon the finding by showing that there was a narrow window in the power-intensity range in which the release of calcium was enhanced. In 1979, they performed a similar experiment with 50 MHz radiation—another ham radio frequency—and found that when its amplitude was modulated at 16 hertz, it also enhanced the release of calcium ions from chick-brain tissue, though at different power intensities from those that had been effective with 147 MHz radiation. In the same year, Blackman teamed up with William T. Joines, an electrical engineer in the Department of Electrical Engineering of Duke University, in nearby Durham, to determine why the two different carrier frequencies required different power intensities to produce the calcium-efflux effect. They found that the two frequencies, though generated at different power, produced identical electric-field intensities within the chick-brain tissue, and concluded that the internal field intensity was what provided the critical parameter for the effect.

The following year, Blackman and Joines studied the influence of a sixteen-hertz field on chick-brain tissue in the absence of a higher-frequency carrier wave. They found that the sixteen-hertz fields increased calcium outflow at two different power intensities, but that fields of one and thirty hertz produced no response. When Blackman reported the findings of his latest study at the third annual meeting of the Bioelectromagnetic Society in August 1981, he noted that they constituted "a solid case of a response that cannot be explained by a general heating mechanism." Later that year, Blackman and Joines extended their low-frequency experiments with chick-brain tissue to cover the range from 1 to 120 hertz. They found a strong response to fields of 15, 45, 75, and 105 hertz, and a much weaker response to fields of 60 and 90 hertz. All of these responses were obtained at power intensities far below those necessary to produce heating.

Ironically, the EPA was now beginning to shut down the research program in the Health Effects Research Laboratory because of budget cuts imposed by the Reagan administration. At the beginning of the decade, there were twenty-seven researchers working on microwave and radio-frequency radiation effects in the laboratory. By 1984, there were only sixteen researchers. And since 1986 almost no funding has been made available by the EPA for research in this area.

Back in 1979, while Blackman was trying to imagine what might account for the unusual ability of radio-frequency radiation modulated at sixteen hertz to alter brain chemistry, he remembered a course in magnetic resonance he had taken as a graduate student at Penn State. "During the course, we had learned that the nucleii of certain atoms and electrons can interact differently with various conditions of high-intensity, static magnetic fields," he said recently. "So I wondered if it might be possible for similar interactions to occur with low-intensity, static magnetic fields, such as the geomagnetic field of the earth. The geomagnetic field generally has a strength of about five hundred milligauss, but it can change strength from place to place—sometimes within a few feet—depending upon such factors as the presence of iron-containing rock, or steel girders in buildings, which can act as conductors of magnetic fields. In fact, the strength of the local geomagnetic field in a given location can range from less than one hundred to more than two thousand milligauss.

"I was on a fishing expedition, of course, but I went down to the local K Mart and bought a couple of hula hoops. Then I went to an electrical supply store and bought some ordinary speaker wire. When I got back to the laboratory, I wrapped the speaker wire in the grooved channels running along the inner circumference of the hula hoops, and made myself a Helmholtz coil—a device named after the nineteenth-century German physicist who invented it—which generates magnetic fields. Before I got around to doing anything with it, however, I decided that the whole business was just a nutball hunch, so I stuck the hoops away in a false ceiling in the laboratory and forgot about them.''

Three years later, Blackman was trying to figure out why fields with frequencies of 15, 45, 75, and 105 hertz should have a strong effect on calcium-ion outflow from chick-brain tissue, while fields of 30, 60 and 90 hertz produced only a weak effect. In the autumn of 1982, he went to James R. Rabinowitz, a physicist at the Health Effects Research Laboratory, and asked for his help in explaining this disparity. Rabinowitz took a book entitled "Introduction to Solid-State Physics" off a bookshelf in his office and showed Blackman an equation for cyclotron resonance—a phenomenon that occurs when a static magnetic field causes moving charged particles to go in circles. The frequency at which this circular motion is produced, Rabinowitz told Blackman, is directly related to the strength of the static magnetic field.

"Suddenly, the lights went on in my head," Blackman recalls. "I went back to my lab, pulled the hula hoops out of the compartment in the false ceiling, and aligned them in a way that would allow me to either increase or decrease the strength of the local geomagnetic field present in the room. Early in 1983, Joines, Rabinowitz, and I conducted a series of experiments to determine whether the ability of a fifteen-hertz field to increase calcium-ion outflow in chick-brain tissue would be altered if we changed the strength of the local magnetic field. And, lo and behold, we found that if we cut the strength of the geomagnetic field in half, the fifteen-hertz field no longer produced the calcium-efflux effect. This introduced a new concept into the mystery of how low-level electromagnetic fields could operate to affect the chemistry of the brain.''

Blackman and his colleagues went on to perform additional experiments showing that the 30-hertz signal they had previously found to be ineffective would enhance calcium outflow if they altered the strength of the local geomagnetic field. In July 1984, Blackman described their latest findings at the annual meeting of the Bioelectromagnetics Society, in Atlanta. He told his audience that the intensity of the local geomagnetic field was an important variable in the calcium-efflux phenomenon, and that the results of the experiments he and his associates had conducted appeared to describe a resonance-like relationship, in which the frequency of the electromagnetic field that can induce changes in calcium-ion outflow was proportional to the strength of the local geomagnetic field. Blackman suggested that the phenomenological findings of these experiments might provide a basis for evaluating the apparent lack of reproducibility of some studies showing that biological effects could be caused by low-level electromagnetic fields. Indeed, he pointed out that if the underlying mechanism of the calcium-efflux response in chick-brain tissue proved to be a universal response in the tissue of other biological systems, including that of human beings, research into the health effects of power-line radiation and other 60-hertz fields in the ambient electromagnetic environment might be subject to unnoticed and uncontrolled influences, depending upon the strength of the static geomagnetic field in the vicinity of the laboratories where the research was being conducted.

As might be expected, a number of prominent scientists in the field were interested. Among them was Abraham Liboff, a physicist at Oakland University in Rochester, Michigan, who, together with Dr. John Thomas and Dr. John Schrot, psychologists at the Naval Medical Research Institute in Bethesda, Maryland, had been conducting a study on the effect of 60-hertz electromagnetic fields on animal behavior for the New York Power Lines Project. Liboff, Thomas, Schrot, and some colleagues had been working for a year without observing any significant results, but after hearing of the findings of Blackman and his associates, Liboff hypothesized that some ions in the brain could be in cyclotron resonance with external magnetic fields, and he and his associates tested his theory by adjusting the local geomagnetic field to 270 milligauss—the level at which lithium ions resonate—and found that rats exposed to 60-hertz fields of less than 100 milligauss exhibited impaired timing discrimination.

Blackman and some of his colleagues have proposed a three-step process similar to Adey's to explain how weak electromagnetic fields can affect calcium-ion outflow in chick-brain tissue. "The first step is the initial conversion of weak electromagnetic energy into a tiny chemical change at a primary reaction site that is at present unknown," Blackman explains. "This reaction site could be a trace metal, or an isotope such as carbon-13, or a biomolecular structure that responds in a resonance-like manner. The second step is the amplification process that has previously been postulated by Adey as occurring at the cell membrane, where the tiny chemical change that has been initiated acts as a trigger, and, through a process known as cooperative transition, produces a kind of biochemical domino effect that amplifies the initial triggering event many times over. The third step is the observable results of this transition, which may be changes in the outflow of calcium ions from the surface of the cell, or intracellular changes in protein kinase and ODC activity.''

In Blackman's own calcium-efflux experiments, the magnetic-field component of low-frequency radiation appears to have played a major role in the hypothetical conversion of electromagnetic energy into physiochemical changes. However, in other experiments undertaken by him and his colleagues, and in studies conducted by other researchers, it has been demonstrated that electric fields alone, as well as electric and magnetic fields in combination, have produced different effects in tissue. ''An example of this is an experiment with chicken eggs that Joines and I and some of our co-workers performed back in 1983 and 1984,'' Blackman says. ''The idea for the experiment occurred to me one day in 1982, when I realized that in all of our experiments over the previous seven years we had used the brains of chickens provided by the Department of Poultry Sciences at North Carolina State University, all of which had been incubated over the usual period of twenty-one days in electrical incubators. Since all of them had been exposed to sixty-hertz electromagnetic fields in the incubators, I wondered if the responses of the brain tissues of the chickens we had used in our experiments might have been altered by such previous exposure. To answer the question, we purchased eggs that had not been incubated and proceeded to incubate one set of them with a sixty-hertz electrical field, and to incubate the other set with a fifty-hertz electrical field—the alternating current that is generated in Europe and much of the rest of the world. The power intensity of the fields that we used to incubate both sets of eggs was ten volts per meter—an intensity that is present in virtually every household in the nation.

''When the chicks hatched, there was no difference between the two sets in terms of numbers of chicks or of abnormalities among them. When we performed our calcium-efflux experiment on the brain tissue of the chicks that had been incubated in the sixty-hertz field, we observed the very same changes in outflow that we had seen in the experiments we had conducted with chickens provided by North Carolina State University. However, when we performed the experiment on the chicks that had been hatched in a fifty-hertz field, we found an entirely different response. The physiological significance of this difference remains unclear, but the findings clearly demonstrate that exposing a developing chick *in ovo* to power-line-frequency electric fields can alter the responses of the brain tissue of the chicken after it has been hatched. What is all the more remarkable about this is that the electric-field intensity of ten volts per meter was measured in air, whereas the internal field strength within the developing chick embryo was millions of times lower."

Could such low-level exposure to power-line fields also affect the developing embryos of other species? "Naysayers like to point out that it is not possible to extrapolate from an avian species to humans, but the fact is, there have been some disturbing experiments to show that humans can also be affected by very weak electromagnetic fields," Blackman replies. "Some of the most important of these experiments were performed back in the late 1960s and early 1970s by Rutgers Wever, a physicist at the Max Planck Institute for Physiology, in Germany. Wever put human volunteers in two underground bunkers that had been shielded against all outside noise and activity. Because the bunkers were underground, the subjects could not know whether it was day or night. The two bunkers were identical in all respects, except that one of them was also shielded from external electric and magnetic fields, and was equipped with facilities for introducing direct-current or alternating-current electric or magnetic fields.

"During the course of his ensuing experiments, Wever found that in the absence of external cues—such as noise, activity, light, and darkness—the human volunteers in both bunkers developed circadian rhythms that did not conform to the customary twenty-four-hour pattern, but ranged anywhere from thirteen to twenty-six hours. Moreover, he discovered that the circadian rhythms which developed among the volunteers in the bunker shielded from external electric and magnetic fields differed markedly from the rhythms developed by people in the unshielded bunker. For this reason, he hypothesized that these differences were caused by the natural electromagnetic fields that were able to penetrate into the unshielded bunker. He then tested his hypothesis by exposing the volunteers in the shielded bunker to direct-current and alternating-current electric fields of ten hertz at power intensities of only two and a half volts per meter—a level that is not only extremely low but virtually ubiquitous in dwellings around the world—and found that the artificially generated alternating-current field affected their circadian rhythms in the very same manner as the natural fields had. This was a remarkable discovery, which, in light of our subsequent findings with chicken eggs, should give us pause for serious thought.''

# THE ELECTRIC BLANKET CONNECTION

ON THE MORNING OF OCTOBER 29, 1985, Dr. Adey and seven other researchers in the field of bioelectromagnetics gathered by invitation at the National Academy of Sciences' headquarters on Constitution Avenue in Washington, D.C. They had come to discuss their work and present their findings at a one-day meeting that was being held by a six-member Advisory Committee on Nonthermal Effects of Nonionizing Radiation, which had been appointed by the National Research Council's Board of Radiation Effects. Among the members of the advisory committee were J. Woodland Hastings, a professor in the Department of Biology at Harvard University, who had presided over the National Academy's committee on Seafarer, in 1976 and 1977, and Dr. Michael L. Shelanski, of the Department of Pharmacology at New York University, who was chairman of the New York State Power Lines Project.

The first presenter was Professor Sol M. Michaelson, the veterinarian from the University of Rochester's School of Medicine and Dentistry, who had testified ten years earlier in behalf of the Rochester Gas & Electric Company that ELF fields from high-voltage lines posed no threat to human health. Michaelson claimed that a large body of scientific literature existed to support the hypothesis that the biological effects of non-ionizing radiation could be explained on a thermal basis, and that no hard evidence existed to explain athermal or low-level effects. He dismissed most of the low-level effects that had been observed by Adey and other scientists, claiming that their physiological significance had not been established, and that with the exception of calcium efflux from chick-brain tissue, many of the findings had not been independently verified. He went on to say that Wertheimer's investigations had been "criticized by engineers for using a false premise that high-current configuration wires can be equated with higher magnetic fields in homes," and that her epidemiology had had the effect of "muddying the waters."

During the discussion period that followed, Adey delivered a scathing denunciation of Michaelson, suggesting that some of the public statements he had made in the past about the safety of microwave radiation were inaccurate. Later in the morning session, Adey reviewed the work that he and his colleagues had performed to demonstrate that weak, ELF-modulated fields could alter brain chemistry and the activity of human lymphocyte and melanoma cells. He also discussed the data indicating that weak fields might act as tumor promoters.

Nancy Wertheimer then gave a comprehensive review of nearly twenty recently conducted epidemiological investigations, which indicated that exposure to electromagnetic fields could result in increased rates of cancer and other adverse health effects in human beings. She said that her own most recent work indicated that pregnancies among couples exposed to the 60-hertz magnetic fields generated by electric blankets and electrically heated waterbeds appeared to be more likely to end in miscarriage than pregnancies among couples who did not heat their beds electrically. She told the members of the advisory committee that fetuses conceived in electrically heated beds had a slightly slower growth rate than other fetuses. She added that similar findings held true for couples whose homes were equipped with ceiling-cable electric heating. She declared that further studies—especially an investigation of the possible association between the use of electric blankets and the development of cancer—were essential.

When the National Academy of Sciences released its advisory committee's report in April 1986, it was apparent that the committee members were not much impressed by what Wertheimer and Adey had told them. They noted at the outset that while investigators had produced an abundance of empirical data to show that non-ionizing radiation was capable of causing low-level effects, the data "fails to provide a consistent picture of the nature of the effects." They went on to say that "In the absence of a scientific conceptual framework that brings meaning to a measured result, the temptation arises to create significance by postulating adverse effects from the interaction of nonionizing radiation with biological tissues and cells." Because much bioelectromagnetic research had been directed at describing the occurrence of adverse health effects, this tended "to lower its credibility in the eyes of some scientists," and they suggested that if the focus of concern were shifted from health effects to basic science, "many of the puzzling questions might find answers."

Wertheimer had conducted her study of electrically heated beds in 1983 and 1984. "I had wanted to do such a study since 1979, because it was apparent to me that electric blankets and heated waterbeds were often major sources of intense, chronic exposure to magnetic fields," she said recently. "When Ed Leeper and I had measured the magnetic fields being given off by various household appliances, we found that although some of these fields were strong in the immediate vicinity of the appliances, they diminished sharply within a short distance of their source. However, when we measured the magnetic field in the vicinity of an electric blanket that had been turned on to any power setting from one to ten, we usually got a reading of between ten and twenty milligauss next to the blanket, and five to ten milligauss six inches away. This, of course, was the field strength to which someone sleeping under the blanket would be exposed."

Wertheimer explained that electric blankets give off relatively high magnetic fields because, although the current that flows through the S pattern of parallel wires in the middle of the blanket is balanced by the current flowing in the opposite direction, thus tending to cancel out the magnetic fields, the current is unbalanced at the outer edges of the blanket. The imbalance causes a significant magnetic field to be generated. "The same holds true for the heating element of a waterbed," she said. "But since it is located on the underside of the bed, the magnetic-field reading we usually got for them was around five milligauss. Considering the fact that children with cancer in our original 1979 study had been exposed to an ambient magnetic field of only about one or two milligauss, while the typical alternating-current magnetic field to which a person in the Denver-Boulder area is exposed is only about one half a milligauss, I was naturally interested in investigating whether the use of electric blankets and waterbeds affected cancer rates. The trouble was that such a study would necessitate extensive interviews with cancer patients, and require considerable time and money to conduct, and I was not in a position to get a grant to finance it. So I had to give up the idea of conducting a cancer investigation. I did, however, find a relatively inexpensive way of looking at the effects that using electric blankets and heated waterbeds might have on fetal development. The idea had come to me when I did the childhood study and found that male children who had died of cancer tended to be either those whose birth addresses had a lower current configuration category than their death addresses, or those with stable addresses who developed cancer after at least one year of postnatal life at a residence near high-current primary wires. In addition, I found that an increased cancer risk could not always be observed among children who were born in a house or apartment with high exposure to magnetic fields—a dwelling, say, within fifty feet of thick primary wires—but could often be observed among both male and female children who moved to such a dwelling after birth. These findings suggested to me that very high exposure to alternating-current magnetic fields was possibly causing many susceptible children to be aborted, which, of course, meant that they could not show up later as cancer cases.''

Wertheimer went on to say that designing a study to show the effect of electric blankets and heated waterbeds on fetal development was a far simpler task than designing a study to determine whether the use of these appliances was associated with cancer. ''The chief reason is that you don't have to select a separate control population. You can compare effects in the same group of people. What you are comparing is the rate of miscarriage among users of electrically heated beds during winter—the season during which electric blankets and heated waterbeds are heavily used—with the rate of miscarriage that occurs among the same people during the summer. In that way, you don't have to worry about possible confounders, such as a couple's pregnancy history, or their smoking, drinking, and dietary habits, because these factors are not likely to make any difference with regard to the season of conception."

Since it is not permissible in Colorado to use vital statistics records to generate a population for a telephone survey, Wertheimer began her electric blanket study by combing Greater Denver newspapers for all birth announcements that had been placed from two Denver-area hospitals in 1982. She found that out of 4,271 families giving birth in the two hospitals during that year, 1,806 (42 percent) had published birth announcements. She was then able to reach 1,318 (73 percent) of these families by telephone and to ask them whether they had used electric blankets or heated waterbeds during the previous eight years, and, if so, when, and at what preferred temperature settings.

After conducting the telephone survey, she obtained routinely recorded data on the 1,256 published births for which she had gathered information concerning parental use of electric blankets and heated waterbeds, and also on 528 of 692 prior births to the same parents since the beginning of 1976. These data included such information as the date of the mother's last menstrual period; the date of birth and the birth weight of the child; and the date of the most recent fetal loss, if any. In order to minimize the possibility of including induced abortions in her study, Wertheimer investigated only those abortions that occurred in the year preceding the conception of live births to the same couples—a method that provided evidence that the parents were not averse to having a child.

When Wertheimer compared the rate of reported miscarriages among users of electric blankets and heated waterbeds from September through June with the rate of miscarriage in this same group of people during July and August—the two months in the Denver region when electric blankets and heated waterbeds are not generally used—she found that the users experienced proportionately more miscarriages during the colder months than in the two summer months, while the rate of miscarriage reported by non-users remained fairly constant throughout the year. She also found that miscarriages among users occurred most frequently between September and the end of January, rather than over the entire period from September through June. Wertheimer believed that the younger the fetus is, the more susceptible it might be to adverse effects from electromagnetic-field exposure, and she knew that most spontaneous abortions are thought to occur before a woman knows that she is pregnant. For this reason, Wertheimer hypothesized that between September and the end of January—a period of increasing cold weather—a susceptible embryo might survive early exposure to the magnetic field of an electric blanket or heated waterbed, when that exposure was relatively low, only to be aborted later, when it was subjected to greater exposure as the weather turned colder and temperature setting on the electric blanket or heated waterbed was turned higher. She further hypothesized that since these comparatively late miscarriages were the ones most easily recognized by the mother, they would also be the ones that ended up being reported on the birth records, whereas many miscarriages that took place between February and the end of June, a period of decreasing cold, would go unrecognized for the simple reason that during this period the highest exposure of the fetus to magnetic fields from electric blankets and heated waterbeds—and thus, presumably, the greatest risk of miscarriage—was likely to occur during the first months after conception.

When Wertheimer and Leeper published the findings of their study in *Bioelectromagnetics* in 1986, they pointed out that since seasonal patterns in the spontaneous abortion rates were seen only in users of electric blankets and heated waterbeds, electrical bed heating ''may have a direct effect on fetal development.'' They added that this effect ''could be due to excessive heat or to electromagnetic field exposure.'' After noting that earlier experimental work had suggested that exposure to electromagnetic fields might cause abnormal fetal development, they said that, ''Whatever the mechanism, it seems important to study the possibility of adverse effects from electric bed heaters further, since their use seems to be increasing as the cost of home heating rises.'' They concluded their paper by suggesting that increased use of electric blankets and electrically heated waterbeds might have contributed to unexplained increase in childhood disability that was reported to have occurred since 1960.

''The results of the electric blanket study greatly increased my confidence in our original hypothesis that alternating-current magnetic fields from high-current electrical distribution wires could have serious health effects,'' Wertheimer said not long ago. ''Until then, I had been nagged by the idea that, in spite of my effort to avoid it, I might have inadvertently introduced bias into my previous studies by failing to blind-code the wiring configuration data. But I had done the electric blanket study completely blind, because when I telephoned the families who had placed birth announcements in the newspapers, I had absolutely no knowledge of whether they'd had a spontaneous abortion or not. The trouble with the electric blanket study was that it contained a built-in confounder that might have influenced its results. This was the fact that, in addition to electromagnetic fields, the blankets generate heat, which, if excessive, is known to have an adverse effect upon sperm. Confounders are what make epidemiology both maddening and fascinating. You can always find them if you look hard enough, and that's what good epidemiology is all about.

"At any rate, the more I thought about the problem, the more I realized that the best way to resolve it was to study the incidence of spontaneous abortion using a source of exposure to alternating-current magnetic fields that weren't likely to generate excess heat. As it happened, I had heard about ceiling-cable electric heating, and learned that, because of the proximity of cheap hydroelectric power, ceiling-cable heating was routinely installed in a great many homes built in Eugene, Oregon, during the 1960s and 1970s. Since ceiling-cable heating is essentially a big electric blanket that has been installed in the ceiling, current flowing at the outer edges of the cable pattern is unbalanced—just as it is in electric blankets—and thus produces quite strong magnetic fields. However, unlike the heat from an electric blanket, the heat generated by a ceiling-cable system is not likely to raise body temperature excessively."

As with her three previous studies, Wertheimer had no financing to help her investigate the cancer effect of ceiling-cable electric heating. Undeflectable as always, she visited the Oregon Center for Health Statistics in Portland, where she obtained the records of those legitimate births for 1983 and 1985 which listed the mothers' addresses as being in Eugene or the neighboring city of Springfield. Among other information, each birth certificate contained the date of the most recent fetal loss, if any, and whether it had occurred before or after twenty weeks of gestation. She then went to the Lane County Assessor's Office in Eugene, and obtained microfiche records giving the type of heat and the address of each single-family home in the Eugene-Springfield postal area. Using telephone books for 1977 through 1984, she traced the families whose birth records she had already obtained in order to discover, when possible, the addresses at which those families had lived at the time of conception and at the time of any previous fetal loss. Finally, she compared those addresses with the assessor's microfiche records to determine which were homes with ceiling-cable heat.

As she had in the electric blanket study, Wertheimer now looked at the seasonal pattern of spontaneous abortion. She observed an increase in fetal loss reported by families living in ceiling-cable-heated homes during the colder months of the year (in Oregon, these include the months of October through May), which did not appear among families living in homes with other types of heating. Moreover, as in the electric blanket study, the excess loss was most apparent in the months of increasing cold weather. For this reason, Wertheimer was able again to hypothesize that although there was probably an increased rate of unrecognized spontaneous abortion caused by magnetic-field exposure throughout the colder months, many more of these abortions were recognized, and thus reported, during the months of increasing cold.

Chapter

# SOMETHING IS HAPPENING

THE FAILURE OF THE National Academy of Sciences either to appreciate the significance of Wertheimer's electric blanket study or to understand the massive public health threat posed by exposure to power-line radiation and other electromagnetic fields was underscored in 1986 by a series of stunning disclosures. In February, the Electric Power Research Institute released the report of a study of the effects of 60-hertz electric fields on three generations of miniature swine, which had been carried out between 1978 and 1981 by Richard D. Phillips and several associates at Battelle Pacific Northwest Laboratories, in Richland, Washington. (Miniature swine had been selected because their body weight approximates that of the average human being.)

Back in 1976, the federal Energy Research and Development Administration (ERDA) had awarded Battelle a multi-million-dollar contract to construct a state-of-the-art exposure system and to replicate the controversial three-generation mouse experiment of Marino and Becker, who had found severe stunting in second- and third-generation mice that had been exposed to 60-hertz electric fields of a strength approximately equal to that which could be expected at ground level beneath a high-voltage transmission line. Although Phillips and his colleagues observed some stunting in their test mice, they were unable to find consistent results throughout the whole range of their experiments. However, in a second study conducted for ERDA to determine whether exposure to 60-hertz electric fields could affect the endocrine function of rats, they found that melatonin content of the exposed animals' pineal glands was depressed. (The pineal gland is known to play a role in regulating circadian rhythm in mammals, and melatonin is a hormone that is involved in this process.)

While Phillips and his co-workers were conducting the three-generation mouse study and the rat endocrine study for ERDA, they were also carrying out the three-generation study on miniature swine for EPRI. After Phillips completed the swine study in 1981, he discussed its findings at the 1982 meeting of the Power Engineering Society in New York City. According to Phillips, EPRI officials viewed these findings as "a major bombshell," however, and had delayed their publication for nearly four years by conducting what he claims was "excessive peer review," and by insisting upon repeated revisions in Battelle's final report. Because he considered these actions unwarranted interference, Phillips left Battelle in October 1984 to take up the post of director of the EPA's Experimental Biology Division, in Research Triangle Park, North Carolina, where Carl Blackman works.

When EPRI finally decided to allow the results of the three-generation swine study to be published, it was easy to see why they might have been considered to be a bombshell. The first- and second-generation progeny of female swine that were born and bred in 60-hertz electric fields of 30,000 volts per meter and then mated with unexposed males not only had lower body weights than the third-generation progeny of unexposed female swine but were afflicted with almost twice as many birth defects. Moreover, during the course of a three-generation rat study that Battelle had conducted for the Department of Energy in the early 1980s, Phillips and his colleagues had observed a threefold increase in birth defects among some of the offspring of female rats that had been chronically exposed to 60-hertz electric fields. When they repeated this experiment, no such effects were observed. The findings of the first rat study, however, paralleled those of the miniature swine study, and the results of both studies provided support for the results and conclusions of the pioneering mouse experiments that had been performed back in 1975 by

Marino and Becker, whose work had been dismissed as worthless by Schwan, Michaelson, Miller, Hastings, and other members of the National Academy's committee on ELF radiation from Project Seafarer.

A few months after EPRI's tardy publication of the results of the miniature swine study, another important revelation about power-line radiation was made at a meeting of the Bioelectromagnetics Society in Madison, Wisconsin. David Savitz, the epidemiologist, now at the University of North Carolina's School of Public Health, who was replicating Wertheimer's childhood cancer study for the New York State Power Lines Project, reported that detailed magnetic-field measurements supported the claim by Wertheimer and Leeper that magnetic-field exposures could be accurately estimated by examining the electrical distribution wiring near a given home. Their finding was confirmed by Richard Stevens, an epidemiologist at Battelle, and some colleagues who were conducting a study for the project to see if leukemia in adults was associated with alternating-current magnetic fields. Indeed, both study groups concluded that magnetic fields were determined by sources outside the house and were not affected by household wiring or power use. This not only provided belated corroboration for what Wertheimer and Leeper had been saying all along, but was also an epitaph for the discredited premise that common household appliances and household wiring were capable of contaminating the home environment with strong magnetic fields.

While Wertheimer and Leeper's wiring configuration concept was being confirmed, further disquieting revelations concerning the apparent link between power-line fields and cancer came from Jerry L. Phillips, a research scientist at the Cancer Therapy and Research Center in San Antonio, Texas. Phillips had conducted a series of experiments showing that cultivated human colon-cancer cells exposed to 60-hertz electric and magnetic fields in combination, and to 60-hertz magnetic fields alone, proliferate more easily and are more resistant to attack by immune-system lymphocytes than unexposed colon-cancer cells. He told *Microwave News* that 60-hertz fields were "capable of producing significant permanent changes in cellular structure and function," and that he considered his experiments to be "very much in concert with the epidemiological studies that have been published."

During the summer of 1986, rumors began circulating among researchers studying the biological effects of non-ionizing radiation that, in addition to confirming Wertheimer's wiring configuration concept, Savitz and his colleagues had developed data to substantiate her finding that the homes of children who developed cancer were situated unduly often near electric distribution lines carrying high current. In the middle of September, three hundred people gathered in Toronto for a week-long International Utility Symposium on the Health Effects of Electric and Magnetic Fields, sponsored by Ontario Hydro and utilities-industry groups. At that time, the utilities industry's confidence in continuing the policy of stonewalling the power-line health problem was severely shaken by the answer that Richard Phillips gave to a hypothetical question from a member of the audience at a jam-packed evening session of the conference: If you had a family with two young children, would you buy a house next to a high-voltage transmission line if that house were $25,000 cheaper than an identical house situated farther away from the line? Phillips replied that he would not buy such a house because he felt that it was probable that health hazards were associated with living so close to a transmission line. That answer was especially troubling to the utilities industry because Phillips, the editor-in-chief of the Bioelectromagnetic Society's journal, had long been thought of as having a conservative and non-controversial attitude toward the possibility that power-line radiation might pose a health hazard. An affable, blue-eyed, and silver-haired man of fifty-nine, with a Ph.D. in physiology from the University of California at Berkeley, Phillips had spent fourteen years at the Naval Research and Development Laboratory in San Francisco, where he studied the biological effects of ionizing radiation. However, his experience with EPRI after he and his colleagues completed their three-generational swine study appears to have convinced him that there was an inordinate reluctance on the part of the utility companies to admit the possibility that 60-hertz fields might prove to be a health hazard. In fact, Phillips says that EPRI "went after Wertheimer because her findings were devastating to the utilities." As for his outspokenness at the Toronto meeting, Phillips explains that he had flown there directly from a meeting on non-ionizing radiation in Freiberg, West Germany, where he had chaired a session on ELF. "I saw the Europeans putting their heads in the sand and ignoring our data as if it didn't exist," he said recently. "Then I flew to Toronto and realized that the utility industry here in North America was going to do the same thing if it got the chance. So when I got the chance, I spoke my mind."

Far and away the greatest blow to the effort of the utility industry to deny that 60-hertz electric and magnetic fields could pose a health hazard came on November 20, 1986, when Savitz and his colleagues announced the results of their long-awaited replication of Wertheimer and Leeper's childhood cancer study: their findings showed that "prolonged exposure to low-level magnetic fields may increase the risk of developing cancer in children." Indeed, Savitz not only found a statistically significant association between all types of childhood cancer and external magnetic fields but also determined that for children in certain high exposure groups—for example, children who lived in homes very close to high-current wires—the risk was more than five times that of the control population.

Savitz's study was all the more convincing because, although it covered the same geographical area as Wertheimer's, it did not include any of the same cancer cases. Moreover, Savitz and his colleagues had taken great pains to investigate all possible confounders, and to answer the major points of criticism that had been leveled at Wertheimer's work. For example, they conducted interviews with as many case and control parents as possible, in order to obtain information about such potentially important factors as family cancer history, residential history, parental occupation history, parental tobacco use and alcohol consumption, exposure of the child *in utero* to X-rays and medication, and illnesses of the mother during pregnancy. They also measured the electric and magnetic fields inside residences that had been occupied prior to the time of diagnosis. In addition, the coding of the wiring configuration at all addresses was done in blind fashion—in other words, by a researcher who was unaware of the case or control status of the occupants.

Since Savitz's study was the second to confirm Wertheimer's findings—the first was the Stockholm study that had been performed by Dr. Lennart Tomenius, and published earlier in the year—it was only natural for other researchers in the field to call attention to the fact that the level of funding for ELF studies had dropped drastically in recent years, and the EPA had removed itself from all research into the health hazard of power-line radiation. As for Wertheimer's response to the news that Savitz had confirmed her earlier findings, she made a simple comment to *Microwave News* that could have served as a fitting rebuke to the National Academy of Sciences: "It is time to accept that something is happening and to find out under what conditions it happens."

At about the same time Savitz announced the results of his study, Richard Stevens and his colleagues at Battelle reported that they had found no link between acute non-lymphocytic leukemia among adults in Washington State and their exposure to ELF electric and magnetic fields at home. Stevens acknowledged, however, that his investigation was weak, because nearly a third of the people who had developed acute non-lymphocytic leukemia were unavailable for study. Moreover, when Wertheimer and Leeper subsequently analyzed his data they found that adults who lived in high-current homes and also used electric blankets, electric mattress pads, or electrically heated waterbeds did in fact have a significantly increased risk of developing non-lymphocytic leukemia, compared with people who did not live in high-current homes or heat their beds electrically. Indeed, Wertheimer and Leeper estimated that, if the association they saw in the data were causal, exposure to electromagnetic fields from power lines and/or electrically heated beds could account for 25 percent or more of this type of leukemia in Washington State.

Chapter

# PROJECT ELF

DURING THE EARLY 1970S, when the Navy was trying to sell Project Sanguine to the residents of Wisconsin, researchers at the Illinois Institute of Technology Research Institute prepared an environmental impact statement for the proposed installation that cited electric blankets as a reason for concluding that Sanguine was safe. "Since the electric field under the blanket is more than 3,000 times the Sanguine field, the induced current flow will be comparable to the current conducted through a person standing near a Sanguine antenna," they wrote. Two sentences later, the IITRI researchers drew what they obviously hoped would be a comforting conclusion: "Since electric blankets have been used widely for many years with no apparent deleterious effects observed, no deleterious effects due to exposure to the weak electric fields associated with a Sanguine antenna are expected."

During the late 1970s, Project Sanguine's scaled-down successor, Project Seafarer, in which the ELF antenna at Clam Lake in the Chequamegon National Forest of Wisconsin would be linked to a similar facility in the Upper Peninsula of Michigan, was turned down by a three-to-one margin by Upper Peninsula voters.

They appeared to be not overly impressed by the National Academy of Sciences report discounting all suggestions that the ELF fields from Seafarer might pose a hazard to human health. In 1979, President Carter deactivated the Clam Lake facility, and in the winter of 1981, the Navy proposed to abandon it. In April 1981, however, President Reagan ordered the Navy to proceed with the construction of a scaled-down version of Seafarer known as Project ELF, which called for the existing Clam Lake antenna system to be made operational, and for fifty-six-mile-long power-line-like antennas to be built near the K. I. Sawyer Air Force Base in Upper Michigan.

Project ELF drew renewed opposition from environmental groups, whose members claimed that further studies were needed to determine whether its electric and magnetic fields could interfere with human biorhythms and cause other adverse biological effects. "We're very worried about the long-term effects on people, plants, and animals," said Bonnie Passamani, of Iron River, Michigan, a leader of the anti-ELF movement. "We think it's a health risk and we don't want it here."

Stanley Kasieta, site manager of the Clam Lake facility for GTE, a division of Sylvania Systems—the firm that had taken over the base and would soon be awarded a $230 million Navy contract to upgrade it—disagreed. "This facility is safer than sleeping under an electric blanket," he said.

In December 1981, Senator William Proxmire of Wisconsin bestowed one of his famous Golden Fleece awards on the Navy for keeping a bull at the Clam Lake facility as part of its ELF health effects research program—something that the admiral in charge of command and control for the office of the Chief of Naval Operations had acknowledged was useless from a scientific point of view. In May 1982, the Navy used the Clam Lake transmitter to send a message to a submarine that was submerged at a depth of 400 feet in the Atlantic Ocean off the Florida coast in order to impress a newspaper publisher, who promptly gave his endorsement to Project ELF. The following November, Wisconsin Governor-elect Anthony Earl confirmed a campaign commitment to oppose the project, and the Wisconsin Natural Resources Board passed a resolution condemning it.

In the spring of 1983, engineers from IITRI reported that they had measured a 76-hertz magnetic field of nearly 1.5 milligauss at an inhabited house a mile from the Clam Lake antennas, and a field of nearly half a milligauss at an inhabited dwelling ten miles away. After analyzing the results of the IITRI survey, Wertheimer and Leeper were quoted in *Microwave News* as saying that the strength of magnetic-field measurement at a distance of ten miles was "surprising." They went on to suggest that "the fields are being dispersed by some means quite different from the usual field dispersal from power lines—perhaps by leaking of ELF currents into power lines and other conducting channels." Since their childhood cancer study had found that homes with an increased cancer risk had an average magnetic field of about two milligauss, they recommended that an investigation of magnetic-field dispersal mechanisms be conducted.

Urged on by Jennifer Speicher and John Stauber, leaders of a movement called Stop Project ELF, the state of Wisconsin filed suit in July 1983 in United States District Court against Secretary of Defense Caspar Weinberger and Secretary of the Navy John Lehman, Jr., to block construction of Project ELF until a new environmental impact statement could be prepared. Wisconsin Attorney General Bronson LaFollette charged that the Navy had failed to revise its 1977 impact statement to take into consideration new studies on the health effects of ELF radiation. During September and October, Federal Judge Barbara B. Crabb presided over a trial and heard testimony from Becker and Wertheimer on behalf of Wisconsin, and from officials of GTE and IITRI on behalf of the Navy. In mid-October, Judge Crabb denied Wisconsin's motion for a preliminary injunction against the ELF system, but warned the Navy that if it began construction on the project before her final ruling, it would do so at its own risk. Meanwhile, the Navy released an environmental impact assessment for the Michigan portion of Project ELF which concluded that "there is no credible evidence of adverse effects to human health or the environment."

On January 31, 1984, Judge Crabb issued a sixty-nine-page ruling that barred the Navy from building the new ELF system, or from upgrading the existing Clam Lake facility, until it had prepared a supplemental environmental impact statement to take into account "the significant new information on biological effects of electromagnetic radiation that has been generated since 1977." Among the post-1977 evidence that Judge Crabb found significant was research on the effects of ELF electric and magnetic fields on cellular function, enzymatic function, animal behavior, and growth and development. Specifically citing some of the findings of Adey and Blackman, the birth defects in test animals reported by researchers at Battelle, and the epidemiological studies that had been conducted by Wertheimer and Leeper, Judge Crabb declared that the Navy had "abused its discretion" in not considering the significant new information that had been developed.

In February 1984, the U.S. Attorney representing the Navy filed a motion asking Judge Crabb to reconsider her ruling on the grounds that "the potential harm to the national defense caused by a delay in implementing Project ELF substantially outweighs any potential environmental effects." At the same time, Secretary Lehman submitted a declaration stating that the Navy's need for an ELF capability was "imperative."

Early in April, Judge Crabb reaffirmed her previous decision and ruled that the Navy could not continue to operate its Clam Lake facility. Toward the end of the month, the United States Court of Appeals for the Seventh Circuit in Chicago upheld her injunction pending completion of the Navy's formal appeal. In August, however, a three-judge panel of the Seventh Circuit overturned Judge Crabb's order requiring a supplemental environmental impact statement for Project ELF. In a two-to-one vote, the appeals court said that the information provided by the plaintiffs in the case "falls short of the threshold of 'significance' at which the duty to prepare a [supplementary environmental impact statement] is triggered," and pointed out, "Were we to require the Navy formally to reassess its proposed action with [supplementary statements] every time some bit of new information appeared, we would be unjustifiably interfering with the Navy's mission."

In their majority opinion, the two appeals court justices did not divulge how they came to dismiss as "some bit of new information" Wertheimer's epidemiological study of childhood cancer and Adey's experimental work on the ability of ELF electromagnetic fields to change the chemistry of the brain and alter cellular function. Later in the month, Wisconsin applied to the Supreme Court for a stay of the appellate court decision, but the application was denied by Justice John Paul Stevens. Meanwhile, the Navy had awarded the American Institute of Biological Sciences (AIBS) of Arlington, Virginia—an organization that claims to represent the interests of biologists in the United States—a $319,000 contract to analyze the post-1977 literature on ELF biological effects for the supplemental environmental impact statement mandated by Judge Crabb. The AIBS selected Professor Hannon B. Graves, a leading consultant for the power utilities industry, to be chairman of an eight-man review committee that was charged with carrying out this task.

In April 1985, the committee issued a 290-page report entitled "Biological and Human Health Effects of Extremely Low Frequency (ELF) Electromagnetic Fields," which concluded that it was "unlikely" the electric and magnetic fields generated by Project ELF would harm people, plants, or animals. None of the committee members and consultants found themselves able to offer any definitive opinions about the capability of ELF fields to cause biological effects. The report's conclusion did not come as a surprise to Robert Becker, who had been invited by Graves to serve on the review committee, and had declined to do so. On August 18, 1984, he wrote a letter to Donald R. Beem, an AIBS staff member who had been designated as project director of the Institute's review of the literature on the biological effects of ELF radiation:

> I have been advised that Dr. Graves is simultaneously engaged in a literature review of the biological effects of 60 Hz for the Florida Electric Power Coordinating Group, a consortium of electric power utilities in that state. The stated aim of this study is to propose standards for population exposure to this frequency. It is my firm belief, based on long involvement with this issue and a continuing extensive literature review, that insufficient data is at hand to render such a judgment.

Becker went on to point out that of thirty-seven people listed as being engaged in the Institute's literature review, he could identify only four who had acknowledged that ELF electric and magnetic fields were potentially hazardous. He also noted that the committee's final report would be based upon resource papers prepared by consultants, with no provision for effective presentation of dissenting opinion. For these reasons, he suggested that the committee would not be able to conduct its business in an unbiased or balanced fashion, and he resigned from it.

The accuracy of Becker's prediction was apparent for all to see when the AIBS committee report was released eight months later.

Meanwhile, thanks to the Seventh Circuit's reversal of Judge Crabb's ruling, opposition to Project ELF melted away. As a result, the Navy began to operate the renovated Clam Lake facility in the summer of 1985, and the Michigan part of Project ELF went into operation in the summer of 1989.

# SAVING THE HONEYBEES

Professor Graves's involvement with power lines in Florida began in 1984. The previous year, as the result of community opposition to the Florida Power & Light Company's construction of a substation in Coral Springs, the Florida State Legislature had authorized the State Department of Environmental Regulation to establish public safety requirements for installing high-voltage transmission lines. In the summer of 1984, the department appointed Graves to be chairman of a five-member Florida Electric and Magnetic Fields Science Advisory Commission which it had set up to review the biological effects of power-line radiation. The Commission was financed to the tune of $200,000 by the Florida Electric Power Coordinating Group, a consortium of state electric utilities.

In March 1985, the Commission issued a 266-page report; like the AIBS report, it concluded that it was "unlikely that human exposure to 60 Hz electric and magnetic fields can lead to public health problems." According to Graves and his colleagues, the only adverse effects associated with high-voltage transmission lines was their potential for interfering with heart pacemakers and for shocking honeybees that might be living in hives beneath the wires. With the latter possibility in mind, they called upon state agencies and electric utility companies to warn beekeepers of the potential risk to their bees, and to inform them that they could eliminate the risk by surrounding their hives with grounded chicken wire.

Seven months later, Graves was retained by Daniel White, the Minister for Health of the Australian state of Victoria, to evaluate the siting of transmission lines by the State Electricity Commission of Victoria (SECV), and to review the health risks of power-line radiation. Graves became involved with Australian power lines in much the same way he had become involved in Florida. In 1983, a committee of the Victoria Parliament had recommended that the SECV run an eight-kilometer-long, 220,000-volt transmission line between the Melbourne suburbs of Brunswick and Richmond. The proposed power line, designed to replace an existing 66,000-volt line, would follow a route through the town of Collingwood and several other heavily populated communities, and cross the grounds of a public school. Public opposition to the line was mounted by several community and neighborhood groups, and in June 1985, a nationwide television program called "Four Corners" broadcast an in-depth report on the hazards of low-level electromagnetic radiation, which included interviews with Adey, Becker, Marino, and Wertheimer. Later that year, residents of Collingwood stepped up their opposition to the line. Meanwhile, the Victoria government brought Graves into the picture in an effort to quell the public furor.

In March 1986, Graves sent White a 119-page report, for which he charged the state of Victoria more than $75,000. Graves concluded that no adverse health effects had been found in humans or laboratory animals exposed to the kind of electric and magnetic fields that could be expected to come from the proposed Brunswick–Richmond transmission line. He added that "a careful review of the literature suggests that such effects are very unlikely." In arriving at this conclusion, Graves resurrected the discredited suggestion that household exposure to magnetic fields was more hazardous than power-line exposure. "The magnitude of magnetic fields next to many common household appliances and electric tools is in excess, often far in excess, of those generated by SECV transmission lines and station environments," he wrote.

Graves went on to criticize Wertheimer's childhood cancer

study, on the familiar grounds that she had not measured actual magnetic fields within the houses she had studied. He then dismissed Milham's study and the dozen or so other epidemiological investigations showing that occupational exposure to 60-hertz electromagnetic fields was associated with an increased risk of cancer, claiming that the workers involved may have developed cancer because of their exposure to other carcinogens. As for Adey's experimental work, "No fundamental equivalent in the whole animal for calcium efflux changes in brain slices has yet been determined." Graves claimed that the three-generation miniature swine study conducted by Richard Phillips and his colleagues at Battelle had failed to verify the stunting found by Marino and Becker in their three-generation mouse study. In addition, he asserted that the epidemiological studies of suicide among people living near power lines which had been conducted by Becker and his associates had not adequately quantified the exposure of their subjects to electric and magnetic fields.

Graves's assertions were countered by Wertheimer, Becker, and Jerry L. Phillips of the Cancer Therapy and Research Center in San Antonio, all of whom were asked by officials of the Collingwood Community Health Centre to comment in writing on his report. Wertheimer said that the report contained "abundant evidence that Dr. Graves is a biased advisor." She was particularly critical of Graves's attempt to revive the old argument about how much magnetic-field exposure someone might receive from household appliances, and his use of IITRI's outdated and erroneous table of measurements to substantiate it. "The user of a hair dryer might, with careful placement, receive a 25 gauss exposure to the tip of his finger on the dryer casing," she wrote. "His elbow would receive drastically less, and his whole-body exposure would be quite modest in intensity, and would, of course, only occur for the rather brief time-period when the hair dryer was in use." Wertheimer noted that in 1984 IITRI had published a corrected version of the erroneous table. "Dr. Graves should certainly have been aware of the correction, and of the misleading nature of the table he has presented to your commission," she wrote.

Robert Becker said that Richard Phillips's three-generation miniature swine study had, in fact, found stunting as well as impaired mating performance among the irradiated animals, and he urged that Graves's inaccurate reference to it be stricken from his report. Becker also pointed out that he and his colleagues had measured electromagnetic field strengths in the vicinity of power lines in one of their suicide studies, had correlated these measurements with the incidence of suicide, and had published their data in *Health Physics,* in 1981. "It is inconceivable that Dr. Graves was unaware of this report in a reputable, peer-reviewed scientific journal, and his failure to mention it in his report raises grave questions as to his objectivity and motives," Becker declared.

Jerry Phillips was also highly critical of some of the statements Graves had made in his report to the Victoria Minister of Health, and in June 1986 he was brought out to Melbourne by the Collingwood Residents Association to testify about the health hazards of power-line radiation. After Phillips stated at a public meeting that magnetic fields from the proposed Brunswick–Richmond transmission line would increase the risk of cancer among people who were exposed to them, the State Electricity Commission of Victoria issued news releases declaring that the claim was "grossly exaggerated and unreal." According to the releases, Phillips was offering a personal opinion that was different from that of Dr. Graves, who was described as a "world authority."

In July 1986, the Victoria Health Department issued a report which concluded that there was no basis for rejecting Graves's findings, and that the electric and magnetic fields from the proposed high-voltage transmission line did not pose any public health hazard. According to officials of the Health Department, "Any charge of bias against Dr. Graves would have to be directed also at the most highly respected scientists working in this field." As for the many studies linking electromagnetic fields with the incidence of cancer, they claimed that "Without exception, each claim has been found unjustified." They went on to say that it was the opinion of Dr. Michael Repacholi, chief scientist at the Royal Adelaide Hospital, and chairman of the World Health Organization task force on ELF fields, that there was not a single confirmed piece of evidence to suggest that standards had to be changed from their present levels.

On August 11, 1986, the Victoria State government granted approval for the construction of the proposed Brunswick–Richmond power line. In a joint announcement, Victoria State health minister Daniel White and Bart Fordham, the Minister for Industry, Technology and Resources, said that the decision was based partly on the fact that no existing epidemiological studies "conclusively demonstrate any health risk associated with high-voltage transmission lines"; that the strength of the electric fields within the proposed Brunswick–Richmond line was one half of the minimum acceptable standard recommended by the World Health Organization; and that "magnetic fields under the transmission lines are extremely low, typically in the range of 5 to 50 milligauss, compared with a WHO minimum standard of 3,000 milligauss." Speaking for the Department of Health, White declared that it was "not likely that future evidence will emerge to show any health risks associated with electromagnetic fields."

Three months after White made this assertion, Savitz announced that he had confirmed the results of Wertheimer's childhood cancer study, and that "prolonged exposure to low-level magnetic fields may increase the risk of developing cancer in children." As a result, opposition to the proposed Brunswick–Richmond line continued to grow, and in March 1988, the government of Victoria ordered that it be reviewed by a panel made up of representatives from the State Electricity Commission, the government, unions, and community groups. At the time, a union official told *Microwave News* that it was "most unlikely that any further overhead system will ever be built in urban areas in Australia."

Meanwhile, back in the United States, the Electric Power Research Institute, which had been busily touting itself as a leading agency for getting to the bottom of the power-line problem, increased its annual budget in March 1987 for studying the biological effects of power-line radiation to $2.7 million, and announced that it had set up a panel of scientific experts to establish priorities for the Institute's senior management. As chairman of the panel, EPRI named Professor Hannon B. Graves.

# 24

# BURIED HEADS

IN SEPTEMBER 1986—two months before Savitz announced that he had confirmed the findings of Wertheimer and Leeper's childhood cancer study—an extraordinary attempt was made to play down the health hazards posed by low-level electromagnetic fields. This occurred when *Scientific American* published an article in its September issue entitled "The Microwave Problem," and written by Professor Arthur Guy of the University of Washington's School of Medicine and Kenneth R. Foster, a professor in the Department of Bioengineering at the University of Pennsylvania. After recounting a history of microwave research that placed heavy emphasis on Herman Schwan's theory of tissue heating through high-intensity exposure, Foster and Guy suggested that public concern was being heightened by unsubstantiated reports that low-level exposure could have adverse biological effects. To prove their point, they launched into a discussion of the three-year rat study that Guy had conducted for the Air Force during the late 1970s and early 1980s. After declaring that the results of the study had "revealed few differences" between the rats that had been exposed to low-level microwave radiation and the non-exposed control rats, they admitted that "one difference was striking; primary tumors developed in eighteen of the exposed animals but in only five of the controls." They then said that although this difference was "statistically highly significant," it should not be interpreted to suggest that low levels of microwave radiation could cause cancer, because, as they put it, "the exposed animals had an excess of tumors only in comparison with the controls, not in comparison with the rate of tumor development generally observed in this strain of animals."

Foster and Guy went on to say that "no single type of tumor predominated," but they added, "If some specific type of tumor had predominated, that finally would have made a much stronger case for a carcinogenic effect from low levels of microwave energy." With this as a rationale, they found themselves able to conclude that "the finding of excess cancer is provocative, but whether it reflects a biological activity of microwave radiation is not certain," and that "although some hazard from weak microwave fields might be proved in the future, there is currently little evidence for the presence of such a hazard."

At this point, Foster and Guy asked "how should future research proceed, and on what basis should any new standards be set?" After suggesting that there should be better coordination of investigations into the biological effects of microwaves, they called for certain restrictions to be placed on further inquiry into the problem, saying that "some criteria must be developed for determining when to halt research on a given topic, open questions notwithstanding."

In its December 1986 issue, *Scientific American* published a letter from Louis Slesin of *Microwave News*. "Contrary to what Guy reports about his own results, the malignant tumors were heavily concentrated in the rats' endocrine systems: nine in the exposed animals compared with only two in the controls," Slesin wrote. He then made a devastating comment. "That a principal investigator for a $4.5-million-study should obscure his own positive results points to a basic feature of the 'microwave problem': the domination of military funding for biomedical research on non-ionizing radiation and the reliance on engineers rather than biologists to do the research." He noted that both Foster and Guy were engineers.

Foster and Guy responded in a letter that was also published in the December issue of *Scientific American*. "The significance of the finding of increased cancer in the irradiated versus control rats is less clear to us than it apparently is to Slesin,'' they commented. ''The study involved 155 comparisons between exposed and unexposed rats. With so many comparisons, seemingly striking differences are expected by chance alone. It is misleading in such case to ignore many negative findings and focus attention on one striking result.''

In a letter that appeared in the November–December 1986 issue of *Microwave News*, Adey and Asher Sheppard declared that the argument advanced by Guy and Foster was untenable. ''Their apology for the elevated cancer incidence has all the aspects of an option more often exercised by politicians,'' Adey and Asher wrote. ''If the facts are unattractive, bury your head in the sand (i.e., statistics) and hope no one will notice the odd posture. Incredibly, from that posture they have the audacity to bemoan the ambiguous nature of the data and call for duplicate studies with an eye to ending research.''

Chapter

# THE FREE EXCHANGE OF THOUGHT

WELL BEFORE GUY, FOSTER, AND PICKARD suggested that further research on the biological effects of low-level electromagnetic radiation should be suspended, the electronics industry had undertaken some initiatives of its own to deal with the problem. In October 1982, the Electronics Industries Association co-sponsored, with the National Association of Broadcasters and the Association of Home Appliance Manufacturers, a three-day seminar on "RF Radiation: Legal and Policy Implications" at The Homestead, in Hot Springs, Virginia. In their letter of invitation the sponsors of the seminar said that the objective of the meeting was to "give participants an authoritative overview of problems caused by arousal of the public's fear of 'radiation,'" and to "address appropriate planning measures aimed at restoring rational attitudes toward nonionizing radiation and electronics equipment associated with such radiation."

The letter went on to say that "One option being examined by the involved industries and users is the formation of an alliance which would operate much the same as the one formed in response to proposals to ban the use of fluorocarbons." This was a reference to the Alliance for Responsible CFC (Chlorofluorocarbon) Policy, which had been organized in 1980 by I. E. DuPont de Nemours, Inc., and other leading manufacturers of chlorofluorocarbons—chemicals then strongly suspected of initiating a chain reaction in the stratosphere that was destroying the ozone layer. In 1978, the EPA had banned the use of CFCs as propellants in aerosol spray cans, but the Alliance lobbied effectively in Congress and elsewhere to head off any further attempts to regulate the chemicals, claiming that the case against them was circumstantial and unproved, and would remain so until actual depletion of the ozone layer could be measured. The effort to extend the presumption of innocence to chlorofluorocarbons met with considerable success—thanks largely to equivocal assessments of the ozone problem on the part of three different National Academy of Sciences committees. In 1985, however, huge and unexpected losses of ozone were measured in the stratosphere above Antarctica, and when chlorofluorocarbons were implicated as the leading culprit, DuPont and other manufacturers announced that they were going to phase the chemicals out of production. Ironically, after helping to delay a solution to the problem for almost ten years, DuPont was given front-page credit by *The New York Times* for implementing a farsighted policy to preserve the ozone layer.

The letter of invitation to the Hot Springs seminar continued: "Today in the United States there is an irrational fear of radiation energy associated with electrical/electronic systems operating at frequencies from 60 Hz to the highest microwave frequencies." Its authors declared that "Allegations of 'radiation' hazards are now applied to high-voltage lines, ELF and VLF transmitters, video-data terminals, broadcast transmitters, microwave ovens, microwave relay dishes, radars, and satellite earth stations." They warned that "the cost to the public and industry in terms of money, time, and unnecessary fear continues to grow as litigation, quasi-litigation and public opposition to electrical technology grows."

Litigation was obviously much on the minds of the organizers of the seminar, who had asked attorneys from the New York Telephone Company, Bell Telephone Laboratories, Pacific Telephone & Telegraph, and Raytheon to hold a panel discussion on ways of improving litigation strategies. Indeed, it was so much on their minds that, after blaming the "radiation fear" phenomenon on irresponsible behavior by a few individuals in both the media and professional communities, they issued a thinly veiled warning about what would happen if such foolishness did not cease: "The general public must learn that the costs of irrational fear of electrical technology can only be borne by the consumer/citizen."

The organizing committee for the Hot Springs seminar had thirteen members, who included its chairman, John Osepchuk, of Raytheon; George A. Kiessling, director of product safety at the RCA Corporation, in Cherry Hill, New Jersey; John Mitchell, of the Air Force; Ronald C. Peterson and Max M. Weiss, of the Bell Telephone Laboratories, who had conducted studies purporting to show that no harmful radiation was being given off by video display terminals (VDTs); and representatives from the Electronics Industries Association, the National Association of Broadcasters, and the Association of Home Appliance Manufacturers. The seriousness with which they took their mission is evident from the last two sentences of the letter of invitation: "To permit a free exchange of thoughts on this sensitive subject, the meeting will not be opened to the press, and no records of the proceedings will be issued. For the same reason, cameras and recording devices will not be permitted in the meeting rooms."

In 1973, Osepchuk, testifying before the Senate Committee on Commerce in behalf of the Association of Home Appliance Manufacturers, had quoted with approval someone describing the hazards of microwave ovens as being "about the same as the likelihood of getting a skin tan from moonlight." At the Hot Springs seminar of 1982, he pushed hard for the formation of an alliance to counter public fear of the health hazards of non-ionizing radiation, and in 1983, he became chairman of the Radiofrequency Radiation Alliance Organizing Committee, which laid plans to present a "vigorous defense in radiation litigation actions, encourage rational government actions, support sound research, and help bring about public education in this area." The groundwork laid by Osepchuk and his colleagues on the organizing committee resulted in the formation of the Electromagnetic Energy Policy Alliance (EEPA) in February 1984. Its founding members were the American Telephone & Telegraph Company (AT&T); GTE; the MCI Telecommunications Corporation; Motorola; the National Association of Broadcasters; RCA; Rockwell/Collins; and Raytheon. By the summer of 1985, EEPA had been joined by CBS, NBC, IBM, and Rockwell International.

Some seventy-five representatives from leading electronics and communications firms across the nation attended EEPA's second annual meeting, held in May 1986, in Washington, D.C. Among their chief preoccupations were how to deal with litigation and public concern over low-level radiation. The first speaker at a session on microwave product-liability lawsuits was Michael M. Futterman, an attorney specializing in medical malpractice, who suggested that lawsuits brought by plaintiffs claiming to have been injured by microwave radiation constituted a "new form of terrorism," which he urged his listeners to combat "with all the means and ingenuity that client and counsel can muster." The next speaker was John Mitchell, who described in detail how the Air Force had used Representative Studds in 1978 to assure Cape Codders that they could depend upon the Air Force to do an honest and accurate job of measuring microwave radiation levels. "We decided that we would go through the congressional office out there since it [the problem] had a lot of congressional attention, and that we would suggest to them that they gather some of their local smart guys, their local citizens, people that the community would have some trust in, and let them go with us and make these measurements," Mitchell said. "As a matter of fact, we said, we will let you pick the spots and you go with us and you check everything we do, and when we're done you give the data to your fellow citizens and see if we can't have a little closure on at least the facts of what the radiation levels are."

Mitchell went on to tell his listeners that the observers had gone back to their fellow citizens and said "Yes, it's true, the Air Force does have a good measurement system." He then boasted about how the Air Force had repeated this maneuver with citizens in California, who were concerned about the potential health hazard of radiation from the PAVE PAWS radar constructed at the Beale Air Force Base, north of Sacramento. "We made the measurements, talked about the data, how it related to the radiation levels, and eventually that did quiet them down," he said, adding that the Air Force was continuing to provide similar assurances to citizens who were worried about the health effects of PAVE PAWS radars about to become operational in Georgia and West Texas. Mitchell said that a question had been raised as to whether it was safe for a local rancher to fly his private plane in and out of a small airstrip located very close to the PAVE PAWS radar in Texas, but he assured his listeners that after making appropriate measurements from a helicopter the Air Force had decided that flying in the vicinity of PAVE PAWS posed no hazard. It was two years after he gave this assurance that the Air Force found it necessary to consider moving the PAVE PAWS at Robins Air Force Base, in Georgia, in order to reduce the danger of explosion in airplanes flying through its main beam.

During the next two years, EEPA continued its self-appointed role of working for what it called "a responsible and rational public policy regarding electromagnetic energy." The featured speaker at the fourth annual meeting, which was held in April 1988, in Alexandria, Virginia, was Dr. David O. Carpenter of the New York State Department of Public Health, who had served as executive secretary of the New York Power Lines Project. Carpenter minced no words as he described the results of the Savitz study to the hundred or so members of his audience, who included representatives from a dozen major utility companies. He told them that Savitz had found that leukemia, lymphoma, and brain tumors—the most common forms of cancer in children—were twice as likely to occur among children living in homes near high-current distribution lines, and that "the basis for the hypothesis that magnetic fields cause cancer is now established." He then suggested that it was not ethical to sit around for another ten years before doing something about the problem, because fully twenty percent of the homes in the Greater Denver area had been found to have elevated magnetic fields. "We are where we were with cigarette smoking and cancer twenty-five years ago," he declared. Research money for studying the problem had "almost totally dried up," and most of the work was now being financed by EPRI, which had a built-in conflict of interest because it was supported by the utilities industry. In conclusion, Carpenter urged that this additional research be financed by federal agencies in an atmosphere "clearly independent of partisan influence."

Carpenter was followed by Edwin L. Carstensen, a professor of electrical engineering, radiation biology, and biophysics at the University of Rochester, who had testified with Schwan, Michaelson, and Miller in behalf of the Rochester Gas & Electric Company at the New York Public Service Commission hearings in 1976. Carstensen said that the strength of the magnetic fields given off by transmission lines was very low, and that a person could be exposed to fields of between twelve and eighteen milligauss simply by virtue of sleeping under an electric blanket. (He did not mention that Wertheimer and Leeper had conducted a study showing that couples who slept under electric blankets experienced a higher rate of spontaneous abortion than couples who did not.) Carstensen went on to say that there was no basis in present knowledge for regulating exposure to magnetic fields, but that there was a "need for good old science to find out if a problem really exists."

Carstensen was succeeded at the podium by Dr. Philip Cole, professor and chairman of the Department of Epidemiology at the University of Alabama's School of Public Health, in Birmingham. A year earlier, Cole had prepared a lengthy critique of Wertheimer and Savitz's childhood cancer studies for the Florida Electric and Magnetic Field Advisory Commission that was chaired by Professor Graves. In this critique, Cole said that Wertheimer's work showed a lack of clarity and logic, an absence of the analytic methods of epidemiology, and an "inordinate consistency." He ridiculed her finding that alternating magnetic fields were associated with an increase in the incidence of all forms of cancer in both children and adults. "I submit that Wertheimer purports to have discovered a universal promoting agent of cancer," he wrote. "If correct, Wertheimer has made the greatest discovery in carcinogenesis in the twentieth century and possibly in all time." As for Savitz's study, Cole acknowledged that it "must be viewed as the definitive work to date," but added that it might also be viewed as "only suggestive of a weak effect."

In a panel discussion that followed the individual presentations, Carpenter, Carstensen, and Cole fielded questions from the floor and argued the problem at greater length. Emphasizing that Savitz had found that a child had almost twice the risk of developing cancer if he or she lived in a home close to high-current wiring, Carpenter pointed out that this could mean that between ten and fifteen percent of all childhood cancer might be caused by exposure to magnetic fields from electrical distribution lines, and that the nation could be facing a major public health problem.

Cole responded by declaring that he was "in the camp of the naysayers," and by pointing out that magnetic fields were ubiquitous. Carstensen said that the strongest bias being brought to bear on the power-line problem was that of individual investigators to find effects, and that there should be no great concern about who was going to finance future studies. Carpenter disagreed strongly, maintaining that the problem being addressed was not just scientific but also political, and that with litigation being filed "all over the place," EPRI should not be in charge of financing biological research.

# THE UNCERTAINTY FACTOR

D<small>R</small>. C<small>ARPENTER'S RESERVATIONS</small> about EPRI can best be evaluated in the light of the Institute's record during the nine years that had elapsed since Wertheimer and Leeper had published the results of their childhood cancer study in the *American Journal of Epidemiology*. To begin with, EPRI made much of the fact that Wertheimer and Leeper had failed to measure indoor magnetic fields in all the homes they had studied. It then used IITRI's erroneous and misleading measurements of magnetic fields given off by household appliances to suggest that Wertheimer and Leeper might be wrong in concluding that high-current electrical distribution lines were the chief source of strong indoor magnetic fields. In 1980, EPRI awarded H. Daniel Roth Associates a contract to reanalyze Wertheimer's data, and in 1981—in an apparent attempt to discredit her work—it sent the New York Power Lines Project a rough draft version of the Roth analysis which claimed that her childhood cancer study was totally invalid. Shortly thereafter, Leonard Sagan, the program manager for EPRI's radiation studies, suggested that "it would be unfortunate for people with children to become overly concerned" about Wertheimer's finding that magnetic fields could cause cancer. In 1983, Sagan told *The New York Times* that "while it does appear that something is going on," he did not believe there was "any reason to alarm the public."

In the summer of 1984, as newspapers around the country were carrying stories about the disturbing results of the rat cancer study that had been performed by Guy and Chou at the University of Washington, the EPRI *Journal* published an article that seemed designed to further play down growing public unrest about power-line radiation. "Electric and magnetic fields surround us," it began. "They are given off by all household appliances, power lines, and even the earth itself. No obvious harm has resulted from nearly half a century of public exposure to the fields from electric power sources and from decades of using electronic equipment. The health effects, if any, that result from normal exposure to these fields are therefore probably very small."

The authors of the article went on to state that between 108,000 and 130,000 miles of new transmission lines would be needed in the United States before the end of the century. They warned that not permitting utilities to use lines carrying 765,000 volts and higher voltages would cost an additional $3 billion. They claimed that EPRI had become "a leading sponsor of research on the biologic effects of electric and magnetic fields," and that "research so far suggests that routine exposure to electromagnetic fields, including those from transmission lines, does not present a public health hazard." However, they carefully avoided defining what they considered a routine exposure to be. Instead, they wrote the following paragraph:

Measuring the intensity of electric and magnetic fields near a power line or appliance is straightforward; estimating the duration of human exposure to particular field intensities is not. For this reason, most earlier studies have not included actual measurements of human exposure to fields. The problem is that people come and go, constantly changing their proximity to the sources of fields. Posture and surroundings can affect exposure—a farmer walking under a power line experiences its field more strongly than one bending over or standing in shoulder-high corn. Frequency also matters. Power lines produce almost exclusively a 60 Hz field, but appliances like TVs and home computers give off harmonic fields containing much higher frequencies. All of these factors have to be taken into account in order to determine whether exposure to fields may affect human health.

The EPRI writers went on to describe an Institute-sponsored study that had measured the annual exposure of farmers tending crops on farms crossed by high-voltage transmission lines. The study found that the domestic exposure of the farmers to electromagnetic fields was greater than their occupational exposure, and that more than half of their domestic exposure was from "using electric blankets, with television sets, light dimmers, and household appliances making up the balance." Moreover, another study was under way to determine more precisely what kinds of electric and magnetic fields people might encounter from various sources in their houses. "Preliminary data indicate that the fields near operating TV sets, home computers, and light dimmers are higher than those directly under nearby distribution lines," EPRI claimed. "Such data cast doubt on previous attempts to estimate field exposures inside a house from proximity to outside distribution lines. Also, the fields emitted by household appliances were found to include higher frequencies that may be more readily absorbed by the human body than the relatively pure 60 Hz field given off by power lines."

After extending the quality of purity to power-line radiation, the authors of the *Journal* article declared that "Such exposure data can help put some recent epidemiologic studies into perspective because none of them has been based on actual measurements of exposure to fields." They then challenged the findings of Wertheimer's childhood cancer study by quoting Sagan. " 'Epidemiologists look at differences in the frequency of a disease between two populations to determine which factors might contribute to disease, but association alone cannot be treated as exclusive evidence of causation,' he said. 'Only after a study has been repeated under a variety of conditions can an epidemiologist conclude that exposure to a particular agent causes a given disease. If the Wertheimer-Leeper conclusions were valid, magnetic fields would be among the most powerful carcinogens known, and there would be an epidemic of childhood cancers.' " (When Wertheimer and Leeper's conclusions were confirmed two years later by the study David Savitz conducted for the New York

Power Lines Project, Sagan would take the position that Savitz's findings were inconclusive.)

After trying to discredit Wertheimer's epidemiology, the au thors of the EPRI article said that the research in cell biology performed by Adey, Blackman, and others "has recently pro- vided useful insights about how fields may affect living organ- isms." They then moved on to the subject of animal studies, maintaining that most of them "have failed to produce noticeable changes in health or bodily function." As for the unpublished, EPRI-financed miniature swine study conducted by Phillips and his colleagues at Battelle, they admitted that there had been a higher incidence of fetal malformation in the litters of piglets re- sulting from the second breeding of the first generation of swine that had been exposed to a 60-hertz electrical field. They de- clared, however, that the results were difficult to interpret, and that a group of three independent teratologists had decided the study had not "conclusively demonstrated the existence of a re- lationship between electric field exposure and fetal malforma- tions."

Concerning EPRI's efforts to repeat the miniature swine exper- iment with rats, the authors admitted that "the second generation of exposed female rats showed a higher incidence of fetal malfor- mations in their second litters than the control group," and that this result "was similar to that observed in the swine." But in a second repetition of the experiment, the same result was not ob- served, they maintained. This allowed them to characterize the findings of the rat study as "ambiguous," and to look for a silver lining in the ambiguity. They did so by quoting Robert Kavet, the EPRI project manager who had sent the rough draft version of Roth's inaccurate analysis of Wertheimer's childhood cancer study to the New York Power Lines Project.

" 'Of course, we would have preferred that the results were unambiguous, but they clearly demonstrate that even exposure to worst-case levels of electric fields—levels perceptible to the ex- perimental animals—cannot produce consistent effects,' Kavet said. 'Obviously, no human mother is going to experience any- thing like this extreme level of exposure, and so our judgment is that the phenomena observed in the laboratory cannot be taken to imply a risk to human health from ordinary encounters with fields from power lines.' " Kavet was, in effect, claiming not to know that in a properly designed developmental study, a relatively small population of pregnant test animals is exposed to a high-level electric field in order to determine if adverse effects can be produced in their offspring; the results of the high exposure levels can then be extrapolated to those that would be produced by the low-level electric fields to which a vast population of human mothers might be exposed.

The EPRI article of 1984 concluded by quoting Rene Males, a vice president of the Institute, who said that EPRI was sponsoring some expensive new animal studies to determine the effects of low-level electric and magnetic fields on reproduction and development. " 'Until then, anecdotal suggestions of possible dangers need to be viewed with a large dose of skepticism,' he declared. 'Current limits on public exposure to electric fields appear to be very conservative, even in light of the most recent research results.' "

Considering that EPRI had been financing studies of the health effects of power-line fields for nearly twenty years, it seemed reasonable to wonder just how serious the Institute was about getting to the bottom of the problem. In 1986, only after Savitz announced that he had confirmed Wertheimer's findings, EPRI awarded a two-year, $350,000 grant to Dr. John Peters, an epidemiologist at the University of Southern California, to investigate the link between all types of childhood leukemia and power-line fields. And in August 1987, EPRI awarded a $300,000 grant for an epidemiologic investigation of utility workers exposed to electric and magnetic fields to Savitz himself.

Interestingly, two months after awarding Savitz the research grant, EPRI finally got around to commenting on his confirmation of Wertheimer's results in an article that appeared in the October–November 1987 issue of the EPRI *Journal*. "The results of a new study are stirring public concern as to whether childhood cancer may be linked to magnetic fields from power lines," the article began. "The findings are highly uncertain, however, and the researchers involved maintain that their work raises important questions but falls far short of offering any conclusions or proof."

The latest *Journal* article undertook to substantiate their opening assertion by quoting Sagan. "This finding is curious," Sagan said. "If the magnetic fields are causing cancer, one would expect a pronounced correlation with the actual field measurements. But that isn't the case. Instead, the stronger correlation is with the pattern of power lines outside the homes. It is conceivable that wiring configuration is a better predictor of long-term exposure than point measurements of the magnetic field because the location of the wires stays the same over periods of years, while instantaneous field levels vary throughout the day. Another possible factor, however, is that some other factor associated with the lines but having nothing to do with magnetic fields may be involved." Sagan went on to suggest that a higher density of power lines and current flow was likely to occur in more crowded, urbanized neighborhoods, where there was more traffic, noise, air pollution, and exposure to hazardous chemicals, and that any of these factors could conceivably contribute to cancer. He did not mention that both Wertheimer and Savitz had taken such potential confounders into consideration in arriving at their conclusions.

The EPRI article then quoted Howard Wachtel, an electrical engineer at the University of Colorado who had helped Savitz conduct his replication of Wertheimer's study. Wachtel said that although the wiring codes seemed to correlate with cancer, whether the cancer was caused by magnetic fields "remains an open question." According to EPRI, Savitz and Wachtel placed different interpretations on the results of their study: "Savitz maintains that a number of factors could be responsible for the cancer cases and that magnetic fields are high on that list of possible agents. Wachtel believes it more likely that some other agent that happens to correlate with the high-density power lines is the culprit."

Ignoring the fact that Savitz is an epidemiologist and that Wachtel is an engineer, and without divulging what either of them believed the other culprit agents might be, EPRI wasted no time in casting doubt upon Savitz's findings. "The fact that these two co-workers interpret the evidence differently highlights the complexity and uncertainty that pervades this field," the article said. EPRI then asked Savitz if, in view of the uncertain risk associated with electric and magnetic fields, individuals and communities should do anything to protect themselves while EPRI-sponsored research was getting to the bottom of the problem. "Probably not," he replied. "There is so much uncertainty that it doesn't seem wise for individuals to avoid a danger that may not exist. If there was a little 59 cent device that would eliminate exposure it would be worth it, but in reality there is so much uncertainty and the costs of reducing exposure are so high that it is difficult to justify such expenditure.''

Savitz's answer seems somewhat puzzling in light of the fact that even as he was being quoted in the EPRI *Journal*, he testified before the House Subcommittee on Water and Power Resources that he had found magnetic fields from electrical distribution wires to be associated with an increased cancer risk for children, and had observed a 40 percent increase in cancer among children living in homes with magnetic-field strengths of more than 2 milligauss. Savitz also said that ''strong associations were seen for leukemia, lymphomas and soft tissue tumors.'' He called for increased federal funding so that a large-scale effort could be mounted to resolve the childhood cancer risk, warning that if the association he and his co-workers had found reflected a causal relationship, and if the exposure to magnetic fields in other parts of the nation turned out to be similar to that in Denver, ''a sizable proportion of childhood cancers would be related to this exposure.''

# ADVERSE WITNESS

MEANWHILE, PROFESSOR EDWIN L. CARSTENSEN, who told the 1988 annual EEPA meeting that there was no need to regulate magnetic-field exposure, had played a major role in a lawsuit that was causing tremendous anxiety among officials of the electric utility industry. This was a case in which the Houston Lighting & Power Company had sued the Klein Independent School District, of Houston, over a 345,000-volt transmission line that crossed school property. The problem had begun in 1981, when the company instituted condemnation proceedings for eight and a half acres of the school district's land, and built a power line that ran within 300 feet of an elementary school, 130 feet from an intermediate school, and less than 250 feet from a high school. From the outset, Donald Collins, the superintendent of the district, had raised questions about the possible health effects of the power line, and when he received no satisfactory answers to his queries the school board had refused to grant Houston Lighting & Power a right-of-way, and the company sued.

The lawsuit went to trial on November 10, 1985, before Judge Edward J. Landry, in Harris County Civil Court, in Houston. On that day, H. Dixon Montague, a 33-year-old attorney with the

Houston law firm of Vinson & Elkins, which represented the school district, called Carstensen, who was the chief expert witness for the Houston Lighting & Power Company, to the stand as an adverse witness for the defendant. (In a condemnation proceeding, the burden of proof is on the property owner, so Montague opened the case for the defense.) Carstensen had been hired only a week earlier by Houston Lighting & Power to replace his colleague at the University of Rochester, Professor Michaelson, who had fallen ill.

Early in his direct examination, Montague asked Carstensen if it was true that he had given testimony on the health effects of transmission lines in approximately twenty previous administrative or judicial hearings.

"That's approximately correct, yes," Carstensen replied.

"Now, in every instance, you were testifying for a power line company?" Montague asked.

"Yes, I believe so," Carstensen said.

"And can you tell the jury what period of time has been spanned from the first time you testified on health effects posed by transmission lines to date?"

"Approximately ten years," said Carstensen, who had testified as a minor witness for the Rochester Gas & Electric Company in the New York Public Service Commission hearings in 1976.

After establishing that Carstensen had a Ph.D. in physics and was not a medical doctor, Montague got him to acknowledge that none of the 116 publications listed in his résumé dealt with the health effects of magnetic fields emitted by transmission lines. He then got Carstensen to acknowledge that he and Charles J. Bennett, the supervising engineer of Houston Lighting & Power's engineering planning division, had gone out to the Klein Independent School District on the Sunday before the trial began to measure the electric and magnetic fields from the power line in question.

"Is it true that the more current that flows from the transmission lines facility the greater the magnetic field emitted by the transmission line facility?" Montague asked.

"That's correct," Carstensen replied.

"Have you suggested to Mr. Bennett, or have you had any desire to go back out to the school complex to make additional field measurements?"

"No."

"Well, you are aware, are you not, Dr. Carstensen, that the magnetic field during the week is probably at its very lowest on Sunday because there's very little use of electricity in our community in comparison to the weekdays?"

"Yes, we compared the magnitude of the current flow on the load on Sunday morning with the highest loads, and it was about one third, roughly, on Sunday morning, of the highest loads that you have during the week," Carstensen replied.

Later in his examination, Montague got Carstensen to admit that he intended to criticize Jerry Phillips's studies concerning the effects of 60-hertz electric and magnetic fields on human cancer cells and on white blood cells, without ever having seen or having tried to acquire any of the data upon which Phillips had based his conclusions.

"Have you called Dr. Phillips in order to get a copy of any papers that he has done in this regard?" Montague asked.

"No," Carstensen said.

"Have you called him in order to get copies of any of the data that he's generated in order to write the papers he has done in this regard?"

"No."

"Have you made any contact with Dr. Phillips at all in order to discuss the conclusions he reached in his studies that electromagnetic fields do cause harmful effects to people?"

"My first meeting with Dr. Phillips was this morning," Carstensen replied.

When Montague questioned Carstensen about Wertheimer's work, Carstensen said that even if children in the intermediate school were exposed to a magnetic field of from six to fifteen milligauss, for seven to eight hours a day over the entire school year he could see "no basis to conclude that they would have a higher risk of cancer than any other population."

After examining Carstensen as an adverse witness, Montague called Professor Marvin L. Chatkoff, an associate professor of engineering at the University of Texas in San Antonio, who had been retained by the Klein Independent School District to make magnetic-field measurements in the district, and to calculate what the increased strength of those fields would be when the 345,000-volt power line became fully operational. Chatkoff told the jurors that he had measured magnetic fields of five milligauss in front of the elementary school and nearly ten milligauss at the intermediate school, and had found a level of more than fifteen milligauss in the school district's parking lot and in front of the high school. He estimated that if Houston Lighting & Power were to activate the transmission line to full capacity, magnetic-field-strength levels at the elementary school would be at least twelve milligauss, and could be as high as twenty-five milligauss at the intermediate school.

Following Chatkoff's testimony, Nancy Wertheimer took the stand and described the results of her childhood cancer study. "Our conclusion was that those children who lived near the kind of wires that put out current and have magnetic fields were two to three more times apt to have cancer than children who lived in ordinary homes," she told the jurors. When Montague asked her whether she believed that children attending schools at the Klein Independent School District would face any risk of developing cancer as a result of being exposed to the magnetic fields that had been measured and estimated by Chatkoff, she said she thought it was "indefensible to expose those children to that experiment." Under cross-examination by Ross Harrison, an attorney with the law firm of Baker & Botts, which was representing the Houston Lighting & Power Company, Wertheimer acknowledged that she had no absolute proof that exposure to electromagnetic fields would cause cancer. "My own belief is it's more likely that it keeps the body from fighting off cancer, or from protecting itself against cancer," she said.

Next, Jerry Phillips took the stand and stated that he had conducted some one hundred and fifty experiments at the Cancer Therapy and Research Center in San Antonio, showing that exposure to 60-hertz electric and magnetic fields increased the growth of cultured human colon- and breast-cancer cells. Phillips also described how other experiments with magnetic fields caused a 40 to 50 percent decrease in the ability of lymphocytes—white blood cells that attack cancer cells and foreign bodies—to undergo the necessary growth and division to perform their task. As for what effect the 345,000-volt power line might have on the health of children attending the Klein Independent School system, Phillips said: "As far as I'm concerned, there is a significant increased risk of cancer development." He added that he could not "understand why the power line was ever placed as it was to expose children unnecessarily to fields that are known to produce biological effects."

When Phillips left the witness stand, Montague called Dr. Harris Busch, an oncologist, who had been chairman of the Department of Pharmacology of the Baylor University College of Medicine in Houston for twenty-five years. A former editor of the distinguished *American Journal of Cancer Research*, Busch placed a metronome on the railing of the witness stand, set it in motion, and said that he would like to use it to explain how a magnetic field might work to either cause or promote cancer. "A metronome is just a device which, in essence, moves in two directions," he told the jurors. "And a sixty-hertz field is a field where first the magnetic wave moves in one direction, and then it moves in another direction, just like this device is doing. Only it happens sixty times a second."

Busch explained that there would be a similar to-and-fro movement on the part of anything magnetic in such a field, and that this to-and-fro movement would occur sixty times a second. "So what this means, then, is that any kind of molecule that is in a person's brain, or in a person's body, is being twisted sixty times a second up and back. Now, we do not function in that kind of an environment. We are earthly people descended from an evolutionary species that's been subject to gravity and the magnetic force of the earth itself. But they have nothing like the power of this movement."

Busch described the relevance of all this to the case at hand. "So we have, then, in this special setting of the Klein School District, imposed on these young children, a set of forces which is twisting and turning the molecules in their brains and bodies sixty times a second," he declared. "Now, the implication that this is a process of concern comes from the epidemiological data which has been accumulating. The future development of those data constitutes an experiment. This is an experiment which is in progress in the children of this district." Busch elaborated on this at some length: "My first conclusion is that an inadvertent prospective experiment is now in progress, testing whether the electromagnetic fields emanating from the 345,000-volt transmission line facility will alter the health of the children of the Klein School District. Now, this experiment literally began when the line was electrified last year. And it is an experiment which was not set up as an experiment, or designed, but it is an experiment which will test the proposition. We will know in due course if they do, indeed, get serious health effects."

Busch went on to tell the jurors that this was no trivial experiment. "It is a daily interaction of the human body and mind with potentially dangerous forces," he said, adding that these forces had been increasingly implicated in recent medical literature in the development of leukemia and other cancers. After reminding the jurors that the schoolchildren would be exposed to strong magnetic fields from the power line for about eight hours each day during the school year, and for periods of up to thirteen years, Busch informed them that nothing could obstruct the passage of electromagnetic fields. "These fields go through glass," he said. "They go through concrete. They simply are not stopped by anything in the environment."

Chapter

# VERDICT

WHEN MONTAGUE RESTED THE case for the defendant, Harrison called as the first witness for the plaintiff Allan B. Hague, a research analyst in Houston Lighting & Power's engineering planning department. Hague testified that two of his five daughters attended the intermediate school, and that he did not believe that the transmission line posed any health hazard. Later, Harrison called Charles Bennett, Hague's superior, who testified that part of his corporate responsibility was to keep abreast of the scientific literature on the health effects of transmission lines, and now said that he had been unable to find any evidence that would lead him to conclude that the lines could have harmful effects. Under cross-examination by Montague, Bennett acknowledged that the Sunday he had measured the magnetic fields at the Klein Independent School District with Carstensen was the first time he had ever made magnetic-field measurements.

After bringing out that the highest measurement Bennett had made that day was eleven milligauss, and that Professor Chatkoff, who had performed his measurements on five separate occasions, had recorded a maximum magnetic-field level of 115 milligauss, Montague asked Bennett if it was true that the current load on the transmission line is greater on a weekday than on a Sunday afternoon.

"No, I think we found when the people get home and turn their air conditioners on—that's when we hit our peak," Bennett replied.

"What was the temperature on Sunday afternoon when you made your measurements?" Montague asked.

"About sixty-five to seventy."

"So they needed their air conditioners on when it was sixty-five or seventy?"

"No."

"It was an overcast day on the Sunday you made your measurements?"

"Yes."

"As a matter of fact, it was raining that day, correct?"

"No, it was raining somewhere in the area," Bennett said. "Some light scattered showers."

"So you chose to bring your magnetic field measurements to this jury on a day when it was overcast between sixty-five and seventy degrees outside, to try to attempt to discredit Mr. Chatkoff, and you made the measurements the day before trial?"

"No, that's not what I did," Bennett protested. "I'm just presenting the facts."

"I pass the witness, Your Honor," Montague said.

When Harrison resumed his direct examination of Bennett, he sought to downplay the magnetic-field hazard from the high-voltage transmission line at the Klein Independent School by showing that strong magnetic fields could also be generated by ordinary low-voltage distribution lines. In so doing, he drew attention to something that the utilities had been trying to keep under wraps ever since Nancy Wertheimer had published her findings on childhood cancer, in 1979—that the chief hazard was not from high-voltage transmission lines but from ordinary low-voltage, high-current neighborhood distribution lines.

"Do some distribution lines put out magnetic fields in such a manner that a person who's exposed to them would be exposed to magnetic fields equal to or similar to magnetic fields put out by the transmission lines?" Harrison asked.

"Yes, sir," Bennett replied.

"You testified earlier that distribution lines are quite often in imbalance," Harrison said. "Is that correct?"

"Yes, sir."

"And imbalance causes magnetic fields to be greater," Harrison said. "Is that correct?"

"Yes, sir."

"What features are there about distribution lines that would cause them to have large magnetic fields?"

" 'Large' in the sense that distribution lines are closer to where people live, where they go to school," Bennett said. "So that's what I mean by 'large.' If a transmission line facility put out, say, more magnetic fields than the distribution facilities, by the time it gets to the actual place where you are concerned, they are probably, you know, the same order of magnitude."

"Distribution lines run through people's backyards?" Harrison inquired.

"Yes," Bennett said.

"Down neighborhood streets?"

"Yes."

"And are located closer to homes than transmission lines might be located to these schools," Harrison said. "Is that correct?"

"Of the eighteen thousand miles of distribution lines that we have in our service area, I would say on average that distribution lines are located closer to homes because that's where they are headed," Bennett replied.

Toward the end of the trial, Harrison called Professor Carstensen to the stand and undertook to repair some of the damage that had been inflicted by Montague during his examination of Carstensen as an adverse witness. Carstensen told the jury that Dr. Busch's theory that cancer might be caused by the rotation of magnetic charges within the molecules of the body was speculative, and that epidemiological studies, such as those that had been conducted by Wertheimer, were subject to "severe limitations and inaccuracies." He declared that his analysis of the experimental data concerning the biological effects of electric and magnetic fields showed that "the bulk of the work that has been done has failed to find effects." When asked by Harrison to describe biological effects that are known to occur from exposure to strong electrical fields, Carstensen said that the tips of leaves at the top of a tree that had been allowed to grow beneath a transmission line might wither, and that honeybees are less productive in their hives when exposed to fields of 4,000 volts per meter or higher, which he attributed to electrical shocks within the hives.

Montague's cross-examination of Carstensen was, if anything, more devastating than his direct examination of him as an adverse witness. This is how it began:

"Dr. Carstensen, if we could, you added one more thing to my 'you were not' list. You are not a medical doctor, correct?"

"Correct," Carstensen said.

"You are not an oncologist?"

"Correct."

"And that's one that studies cancer?"

"Correct."

"You are not an epidemiologist?"

"Correct."

"You are not a cell biologist?"

"A great deal of my—"

"You are not at—"

"A great deal of my professional career has been involved in cell biology," Carstensen said. "Probably ten years of intensive research."

"Well, when you were on the stand earlier, Dr. Carstensen, you said you weren't a cell biologist, but now you're saying you are a cell biologist because you have worked with plant cells, primarily, in the last ten years. Is that correct?"

"I have worked with plants," Carstensen replied. "I have worked with micro-organisms. A major part of my dielectric work had been involved with studies of bacteria, yeast, and that sort of thing."

"Bacteria and yeast?"

"Correct."

"Now, that doesn't have anything to do with a human cell, does it?"

"These are not human cells," Carstensen said. "That's correct."

"Nor are they anything like animal cells?"

"My Ph.D. thesis was done on human cells."

"I'm not asking what your Ph.D. thesis was done on," Montague told him. "I'm asking specifically if you have had any formal training as a cell biologist."

"No," Carstensen said.

"You aren't an epidemiologist either, are you, Dr. Carstensen?"

"Correct, I am not."

"And the final thing that you added today, or, at least, which is on my list—you aren't a statistician?"

"That is correct," Carstensen said.

"But you feel qualified to come to these jurors today and comment on the work done by an epidemiologist who has testified in this trial, correct?"

"Correct," Carstensen said.

"Qualified to testify about the opinions and work done by an oncologist and a biochemist, correct?"

"Correct."

"You feel qualified to come before these jurors and testify to the opinion reached by Harris Busch, who is a pharmacologist, oncologist, and cell biologist, correct?"

"I believe that's what he testified to."

"Well, that's not the question, Dr. Carstensen. I asked you if you felt—you, yourself—qualified to come to these jurors and tell them in your opinion that Dr. Busch didn't know what he was talking about?"

"In the specific postulate which he gave, I do feel qualified, yes," Carstensen replied.

After Montague and Harrison presented their final arguments, Judge Landry delivered his charge to the jury, instructing the jurors to answer seven questions on specific issues relating to the case. The jury retired and deliberated for about four hours, before returning with its verdict. To begin with, the jury found that the Houston Lighting & Power Company had "abused its discretion" in taking approximately eight and a half acres of the Klein Independent School District's property for its power line, and that this constituted a "willful and unreasoning action." The jury proceeded to award the school district $104,275 as compensation for actual damages that it had sustained as the result of the utility's use of school district land for some three and a half years. The jurors also found that in constructing and operating the power line on school property, Houston Lighting & Power had acted with "reckless disregard" and with "conscious indifference to the rights and welfare of the persons affected." As a result, they awarded the school district $25 million in punitive damages "as an example that [the utility's] conduct will not be tolerated." The jury went on to find that the reasonable cost to the school district of replacing or restoring its property and facilities to their original condition would be $42,113,120. (What the jurors had been asked to decide was how much compensation the school district was entitled to receive if Houston Lighting & Power had not acted with reckless disregard, and were allowed to keep its power line on school property; their answer was, in effect, that it would cost somewhat more than $42 million for the school district to move its facilities to a new site.)

Given the jury's verdict, Judge Landry entered a mandatory injunction ordering Houston Lighting & Power to shut down its power line and to remove it from school property. The utility quickly asked a three-judge Court of Appeals for permission to keep the power line in operation while the rest of the case was on appeal. Early in 1986, the Appeals Court told Judge Landry that he must allow the utility to turn the line back on. Montague appealed this ruling to the Texas Supreme Court, and at the end of 1986, the Supreme Court unanimously overruled the Appeals Court by upholding Judge Landry's injunction to shut the power line down.

A year later—in November 1987—the Appeals Court reversed the $25 million punitive damage award, on the grounds that when Houston Lighting & Power took school district land for its power line, it had been in technical compliance with the Texas Property Code, and, since it was not guilty of trespass, it could not be held liable for punitive damages. (Montague appealed this decision to the Texas Supreme Court, where it was turned down in May of 1989.) At the same time, the Appeals Court judges upheld the rest of the trial jury's verdict, ruling that the jury had been correct in finding "clear and convincing evidence" of potential health hazards caused by electromagnetic fields, and in concluding that Houston Lighting & Power had abused its discretion in siting a 345,000-volt power line on school property. Meanwhile, during the summer of 1987, Houston Lighting & Power had removed the power line from school property and rerouted it at a cost of more than $8.5 million. Since then, the members of a family whose land had been condemned by Houston Lighting & Power for the right-of-way of the same 345,000-volt transmission line have sued the utility, alleging that electromagnetic fields from the line were responsible for a brain tumor that one of them developed in 1987.

# NOT READILY
# TRANSLATABLE

ON JULY 1, 1987, the scientific advisory panel of the New York Power Lines Project issued a final report of the five-year, $5 million research program that had been undertaken as a result of an agreement between the New York State Public Service Commission (PSC) and the Power Authority of the State of New York, following power line hearings that ended in 1977. The report was entitled "Biological Effects of Power Line Fields," and its authors included Dr. Shelanski, who had now become chairman of the Department of Pathology at the Columbia-Presbyterian Medical Center in New York; Ernest Albert, a professor of anatomy at the George Washington University Medical Center in Washington, D.C.; Antony Fraser-Smith, a research scientist in the Department of Electrical Engineering at Stanford University; Alan J. Grodzinsky, a professor of electrical and bioengineering at the Massachusetts Institute of Technology; Michael T. Marron, a chemist in the Molecular Biology Program at the Office of Naval Research in Arlington, Virginia; Alice O. Martin, a professor of obstetrics and gynecology at the Northwestern University Medical School in Chicago; Michael A. Persinger, a professor of psychology at Laurentian University in Sudbury, Ontario; Dr.

Edward R. Wolpow, a professor of neurology at the Harvard Medical School; and Anders Ahlbom, an epidemiologist in the Department of Epidemiology at the National Institutes of Environmental Medicine in Stockholm, Sweden.

"Biological Effects of Power Line Fields" was poorly organized, equivocal in its assessments, and contradictory in its conclusions; it was apparently designed to persuade the public that 60-hertz radiation could cause only a few effects and that the significance of these effects would have to await further study. The report's flaws emerged in the introductory "Non-Technical Summary." The lead paragraph declared that most of the sixteen research groups in the study program had reported "no effects of concern." (The fact was that no fewer than twelve of the seventeen projects in the research program had reported statistically significant biological effects, but the members of the scientific advisory panel did not get around to acknowledging this until much later in their report.) The second paragraph said that there were some "small but consistent" changes in behavior and brain function, and that changes had been observed in the ability of rats to learn. The third—and final—paragraph read as follows:

> A more serious concern comes from a study of cancer suggesting that children with leukemia and brain cancer are more likely to live in homes where there are elevated 60 Hz magnetic field levels than are children who do not have cancer. Although much more research is needed before the question whether the magnetic fields actually cause or promote cancer can be resolved, the basis for such an hypothesis is now established. At this time no risk assessments can be made because only four studies of this question have been made and the two which report an association are from the same geographical region. More research on cancer as a function of magnetic fields is needed, both in homes and for on-the-job exposure.

The fact was that Savitz had conducted his investigation in the same geographical region—Greater Denver—in which Werthheimer and Leeper had conducted their study because he had been specifically urged to do so by the members of the scientific advisory panel, who had come to the belated realization that Wertheimer's findings were far too important and disturbing not to warrant replication. The suggestion of the panel members that the common denominator of the two epidemiological studies might be Greater Denver, rather than chronic exposure to alternating magnetic fields, set a tone that pervaded the rest of the report. Indeed, they managed to construct a carefully balanced mobile in which each positive finding was counterbalanced with a negative result, or some possibly equivocating factor. After declaring that Marino's finding of stunted growth in rats and mice exposed to 60-hertz electric fields had not been confirmed, they tried to diminish the importance of the increased incidence of fetal malformation observed in the three-generation swine study conducted by Phillips and his colleagues at Battelle by pointing out that the researchers had said that they could not "unequivocally conclude that there was a cause-and-effect relationship." As for the calcium-efflux effect that had been discovered by Bawin and Adey, and studied by Blackman and his colleagues at the EPA, they chose to emphasize differing results attributable to different experimental techniques, and to disregard the essential fact that both research groups had found that exposure to weak fields could alter brain chemistry.

When it came to the childhood cancer findings, the members of the scientific advisory panel could not escape drawing at least one disturbing conclusion. At the bottom of page 85 of their report, they concluded that if the wiring code distribution that had been observed by Savitz in Denver proved to be valid for other locations in the nation, and if the assumption that magnetic fields cause cancer in children held true, "this would mean that 10–15% of all childhood cancer cases are attributable to magnetic fields." In the final section of their report, however, they performed an intricate dance of avoidance on the hot coals of this astonishing possibility. "Research sponsored by this project and related research has demonstrated a variety of effects of electrical and magnetic fields," they wrote. "These findings do not readily translate into concrete regulatory recommendations on width of right-of-way, line heights, or location of lines near homes." After providing themselves with an excuse for not recommending preventive measures, they urged that further research be conducted on methods of delivering power that would reduce magnetic-field exposure; on the interactive effects between the earth's geomagnetic field and 60-hertz fields; on the ability of magnetic fields to affect learning ability; and on the association of magnetic-field exposure and the development of cancer.

In refusing to draw any conclusions from the research of the Power Lines Project, the members of the scientific advisory panel were leaving the unpleasant task of recommending preventive measures to a six-member Power Lines Project evaluation task force, which was made up of staff members of the Public Service Commission. On January 11, 1988, after studying the problem for several months, the task force sent the Public Service Commissioners a twenty-nine-page report analyzing the scientific advisory panel's final report.

At the beginning of the new report, the task force members took dutiful note of the PSC's ten-year-old decision that the strength of the electric field at the edge of all future transmission-line rights-of-way should not exceed 1,600 volts per meter, and that there would be a moratorium on higher electric fields until the results of the Power Lines Project research programs could be evaluated. They also noted that in 1978 the commissioners had selected the interim 1,600-volt-per-meter limit so that "the risks, if any, of long-term exposure to transmission-line electric fields would be no greater than those which society had implicitly accepted of long-term exposure to the 345 kV lines operating for many years throughout the state." They informed the commissioners that fifteen of the seventeen studies in the project had used electric and magnetic fields whose strength was within about one order of magnitude—a factor of ten higher or lower—of the maximum fields typically found near ground level on transmission-line rights-of-way. (The maximum electric field is about 10,000 volts per meter and the maximum magnetic-field strength is approximately 700 milligauss.) They then told the commissioners that magnetic-field exposures in the two epidemiological studies—those conducted by Savitz and by Stevens— were only about 2.5 milligauss, because these studies were chiefly concerned not with high-voltage transmission lines but with the much lower voltage electric distribution lines that run through city streets and suburban neighborhoods. Under the heading "Project Results," they passed on the equivocal evaluation that had been made by the members of the scientific advisory panel:

Twelve of the 17 projects reported statistically significant biological effects under certain sets of laboratory test conditions. However, neither the researchers nor the Panel identified any adverse human health effects from the laboratory studies. Although the sponsored research and other research reviewed by the Panel demonstrated a variety of subtle effects of electric and magnetic fields, the Panel was unable to make any concrete regulatory recommendations on width of rights-of-way, line heights, or location of lines near homes. Because most of the experiments used high field levels typical of transmission lines rights-of-way, most researchers were not able to examine whether an effect persisted at lower field intensities or whether there was a threshold below which an effect disappeared.

The last sentence was scientifically suspect for two reasons. First of all, high levels of exposure—whether to toxic chemicals, asbestos, or radiation—are used in animal experiments to increase the possibility that biological effects, if they are to occur, will occur in the relatively few rats, rabbits, or guinea pigs that are available for study. The high-dose observations in the test animals can then be extrapolated mathematically to equate the lower-dose levels to which a huge human population may be exposed. Second, the fact that most researchers were not able to determine whether the effect persisted at lower levels was irrelevant, because, as a simple matter of prudent public health policy, the panel and the task force should have assumed that the effect *would* persist at lower levels until it had been conclusively demonstrated that it did not. A further absurdity of the panel's caveat was revealed two pages later, when the authors of the task force report repeated the estimate that ten to fifteen percent of all childhood cancer cases might be attributable to the magnetic fields found in many homes.

The six members of the task force did not venture to say whether they believed that parents had "implicitly accepted" the extraordinary potential risk of childhood cancer—a risk associated with exposure to magnetic fields of only two and a half milligauss—in the same way that they were supposed to have recognized and accepted the hazard of long-term exposure to magnetic fields in the hundreds of milligauss from 345,000-volt power lines. Instead, they chose to cite a form letter that Dr. Carpenter, the executive secretary of the Power Lines Project, had been sending out in response to inquiries about the Savitz study. In his letter, Carpenter pointed out that children who live in homes where parents smoke face a far greater risk of developing cancer from that exposure than they do from the effects of magnetic fields. Invoking parental cigarette smoking to diminish concern about childhood exposure to alternating-current magnetic fields is hardly scientific, not the least of the reasons being that children exposed to magnetic fields are prone to develop leukemia, lymphoma, and cancer of the brain (as Carpenter himself had pointed out at the EEPA meeting), not lung cancer.

In the end, the members of the task force avoided making any recommendations that might place the commissioners in the awkward position of having to consider meaningful remedial action. They urged the commissioners to "encourage the federal government to undertake relevant research" and to "plan for an epidemiological study of childhood cancer and magnetic-field exposure in New York State." They also recommended that the commissioners direct the utilities to support research on methods of power delivery that would reduce magnetic-field exposure, thus suggesting that the utilities be the ones to survey the extent to which people were being exposed to electric and magnetic fields from overhead and underground power lines. It is not known what kind of legal advice was available to the members of the task force, but any attorney with knowledge of product-liability law could have told them that if magnetic-field exposure is proved to be carcinogenic, the utilities who supply electric power may be liable to judgment under the law for their failure to warn the consumers of its disease-producing potential.

Finally, the task force members advised the commissioners to continue the ten-year-old interim electric-field standard of 1,600 volts per meter at the edge of transmission-line rights-of-way, and recommended that they "adopt an interim edge of right-of-way magnetic field standard to ensure that future transmission lines do not produce magnetic fields greater than the fields typical of the many existing 345 kV lines operating throughout the state." They then undertook to defend this recommendation with this tortured paragraph:

> Central in the minds of individuals familiar with the Power Lines Project is the epidemiological study of childhood cancer. Clearly, if

this research had established a causal link between cancer and magnetic field exposure, a careful risk assessment (to put the increased risk of cancer in proper perspective compared to other risks society tolerates) and appropriate regulatory action would be required. Further, because the magnetic fields associated with the increased cancer risk (on the order of 2.5 milligauss) are equivalent to those found in many homes, the appropriate regulatory response could involve fundamental changes in the way electrical energy is distributed and used in society. But a causal link has not been established. Additional research in this area is needed.

In the last three pages of their report, the task force members performed a variation of the dance of denial that had been staged by the scientific advisory panel, whose choreography they were evaluating. They not only concurred with the panel's conclusion that the findings of the Power Lines Project's research program "do not readily translate into concrete regulatory recommendations on width of right-of-way, line heights, or location of lines near homes," but declared that "research revealed no evidence that magnetic fields pose a health hazard." This highly dubious assertion paved the way for the final two paragraphs of their report. The first paragraph read as follows:

An interim magnetic field standard should ensure that magnetic fields at the edge of future transmission line rights-of-way are no greater than the fields typical of the many existing 345 kV lines operating throughout the State. Based on a limited number of measurements of transmission lines in New York State, the proper interim limit may be about 100 milligauss. However, if the Commission chooses to adopt an interim magnetic field standard, additional information about existing transmission lines would be needed to determine the proper interim magnetic field limit. Since the limit would not be directly related to biological effects, information about the effects of magnetic fields would not be required.

Because the last sentence of this paragraph might cause anyone to wonder what an interim magnetic-field limit could possibly be related to, if not to the biological effects of magnetic fields, the task force members gave some final words of advice to the Public Service Commissioners, which brought them full circle to the conclusion that the commissioners themselves had reached ten years earlier:

If a magnetic field limit is adopted, it should be made clear that magnetic fields have not been shown to be hazardous and that the purpose of the limit is to ensure that exposures to magnetic fields in future transmission line designs would be no greater than those which society has implicitly accepted for the 345 kV lines operating for many years throughout New York State.

On February 24, 1988, the Public Service Commissioners met in Albany and issued an order continuing the ten-year-old transmission-line electric-field standard, and directing that an interim magnetic-field standard be developed. They noted that their staff had suggested an interim limit of 100 milligauss, but said that additional information about the magnetic-field strength of existing transmission lines was needed before the standard could be set. As a result, they announced that they would convene a technical conference and direct New York State electric utilities to provide data concerning the magnetic fields being generated by the state's 345,000-volt transmission lines. Their next sentence indicated that they, too, had decided to participate in the ballet of the absurd: "Other parties will be welcome to participate in this conference, but since the interim standard will not be directly related to biological effects, information about the effects of magnetic fields will not be considered at the conference." As if concerned that they might be accused of failing to protect the civil rights of magnetic fields, they added that "although we are considering the option of an interim magnetic field standard, we emphasize that magnetic fields have not been shown to be hazardous, and that the purpose of any standard would be to ensure that exposures to magnetic fields in future transmission line designs would be no greater than those which now exist for the many 345 kV lines operating for years throughout New York State."

What would have been laughable about such posturing were its consequences not so potentially disastrous is that the interim standard the commissioners were considering was not only thirty to fifty times greater than the magnetic-field levels demonstrated to be associated with a twofold increase in the development of fatal cancer in children, but also far greater than the alternating-current magnetic fields being generated by thousands upon thousands of miles of ordinary low-voltage distribution lines throughout New York State. Indeed, such fields undoubtedly exist in many homes, residential neighborhoods, school buildings, and business districts from one end of the state to the other.

# TESTIMONY

WHILE THE NEW YORK PUBLIC SERVICE COMMISSION was failing to deal with the power-line problem, Congress did take a belated step toward coming to grips with it. On October 6, 1987, the House Subcommittee on Water and Power Resources, whose chairman was Representative George Miller, a Democrat from California, held a one-day hearing on the health risks associated with the electromagnetic fields from power lines. The first witness to testify at the hearing was Dr. Robert Becker, who nearly fourteen years earlier had warned the New York Public Service Commission that Navy research showing that ELF radiation could cause adverse biological effects might have profound implications for people exposed to electric and magnetic fields from power lines.

Becker told the subcommittee members that man-made electromagnetic fields constituted a health hazard to much of the population of the United States. He stressed that when he had begun his own research back in 1958, the accepted scientific doctrine was that non-ionizing electromagnetic energy had no biological effect on living organisms except the ability to heat tissue at high power. "We now know that very small electrical currents are

generated by living organisms and that they are important regu-
lators of growth and the operations of the central nervous system," Becker said. "Similarly, it has been discovered that the
brain itself, as well as other organs, actually produce magnetic
fields detectable outside of the body. Within the past decade, we
have found that living organisms have specific organs, developed
very early in evolution, whose jobs it is to sense the changes that
occur in the earth's magnetic field and alter the organisms' behavior appropriately. Finally, only within the last year have we begun
to understand the actual physical mechanisms involved in these
interactions between very small magnetic fields and living things.
This area has grown from one in 1960, when less than a handful
of scientists was involved, to one today when we have three
accredited scientific societies, two scientific journals devoted to
this discipline in this country alone, and literally thousands of
scientists involved on a worldwide basis. The question no longer
is one of 'Do very small electromagnetic forces have any bioeffects?' but 'What is the level of hazard from abnormal electromagnetic energy?' "

Becker explained that living organisms had evolved over the
past three billion years in an electromagnetic environment that
consisted primarily of the earth's magnetic field. "Since 1900, we
have markedly changed this environment with the introduction of
fields and frequencies that never before existed on earth. This use
of electromagnetic energy for power and communication has
markedly accelerated since the end of World War II, and we have
now just about filled up the available space in the electromagnetic
spectrum. This change in our natural environment is actually the
most drastic alteration made by mankind and is far greater than
any chemical contamination yet recorded. This was done in the
complete confidence, based upon the 'Thermal Effects Only'
dogma, that no biological effects or actual harm to living things
could occur. We now know that this was wrong."

Becker said that by impairing the immune system and altering
normal cell division, electromagnetic fields could already be having a widespread effect upon human health. "It is instructive to
look at the recent statistics for these general disease patterns in
our society," he told the committee members. "The incidence of
birth defects has approximately doubled over the past twenty-five
years. The incidence of cancer in general is increasing approximately one percent per year, and certain types have become epi-

demic in nature. In 1984, the National Institute for Mental Health reported that the incidence of serious mental disorder in the general population was twenty percent, with the incidence in the under-forty-five age group at twice that figure. The incidence of suicide in the teenage group has more than doubled between the years 1961 to 1981 and while firm data is not available, this increase seems to be accelerating. It is recognized that our society contains other factors that may contribute to this situation. However, the link with abnormal electromagnetic fields is the only one that extends globally to all of these conditions." Becker concluded his testimony by calling for the establishment of a congressionally mandated research program, which would be overseen by a panel of outside experts drawn from many disciplines, and would include scientists who believed that ELF radiation could have harmful consequences. "Full and prompt disclosure of all data and panel deliberations must be assured," he declared.

Becker was followed to the witness table by Jerry Phillips, who told the committee that electromagnetic fields could suppress immune-system function in two ways: by affecting white blood cells directly and by acting as a general source of stress. "If a link is demonstrated between exposure to electromagnetic fields and adverse effects on human health, the potential liability faced by utilities, land developers, and municipalities will be staggering unless a conscious effort is made to protect the public," Phillips said.

Next, David Savitz told the subcommittee about the results of his childhood cancer study, and urged that the Congress support a major research program. Dr. Ross Adey then informed the committee members that exposure to such fields can also suppress the ability of the immune system to function properly by disrupting and distorting cellular communication, and can thus lead to the development of cancer. Like Becker, Savitz, and Phillips, he called for an adequately financed research program that would be conducted by teams of appropriately skilled scientists.

The subcommittee members then heard testimony from Leonard Sagan, manager of the Electric Power Research Institute's radiation studies program, who appeared at the hearing with Dr. Philip Cole, the chairman of the University of Alabama's School of Public Health. Sagan started out by explaining that EPRI, with financial support from the utility industry, had been conducting research on the biological effects of electrical fields from power lines for more than a decade, and that it now intended to conduct animal studies in order to determine if magnetic fields promote cancer. According to Sagan, Savitz's replication of Wertheimer's childhood cancer study was not clear cut or conclusive. "All we know about the exposure of those children was the size of the lines passing the house," he said. "We don't know about electric blankets, or hair blowers, or what they experienced at school or elsewhere. So there is a good deal of uncertainty in our minds."

In addition to uncertainty, there was also apparently a good deal of apprehension in the minds of Sagan and his colleagues at EPRI. Ten months earlier, an EPRI vice president was said to have described the debate over the health effects of power-line radiation as "a jugular issue for the industry," and Sagan himself had been quoted in the December 3, 1986, issue of *The Energy Daily* as indicating that the link between transmission lines and cancer was worrying the industry and needed to be resolved. "Lawyers know about this stuff," he told the paper. "Reporters will know about it soon enough. We have no choice."

When the hearing was over, Congressman Miller wrote to Sagan and other witnesses, asking them to follow up their testimony by supplying written answers to a number of detailed questions. "In 1979, Dr. Nancy Wertheimer found results which some considered startling," Miller wrote in his letter to Sagan. "Why didn't EPRI agree to fund a repeat of the Wertheimer study to confirm or deny her findings?"

Sagan's response seemed both evasive and misleading. "EPRI was not asked to fund a repeat of the Wertheimer study," he maintained. "We chose not to take an initiative to repeat this work because the New York Power Lines Project undertook a replication, the results of which are now available." Sagan not only failed to tell Miller that EPRI had attempted to discredit Wertheimer's study, he also neglected to mention that back in 1982 EPRI had sent a highly critical (and, it turned out, inaccurate) assessment of her findings to the scientific research coordinator of the New York Power Lines Project, whose science advisory panel was in the process of deciding whether to replicate her work.

A few weeks later, Sagan was asked by a newspaper reporter to comment on the studies showing that exposure to 60-hertz electromagnetic fields was associated with the development of cancer in children. "I think it would be unfortunate for people with children to become overly concerned about this," he said. "The evidence is weak and inconsistent, and in a few years we will have better evidence."

During 1988, an important legal battle over power lines began in Goshen, New York, where 58 landowners had filed a $66.5 million lawsuit in the New York State Court of Claims against the New York Power Authority, as the Power Authority of the State of New York was now called. The plaintiffs in the case alleged that the 207-mile-long, 345,000-volt transmission line called the Marcy South Line, which the Power Authority had built from Marcy, in Oneida County, to East Fishkill, in Dutchess County, and had gone into operation in May 1988, had created a "cancer corridor" that had destroyed the market value of the land adjacent to its right-of-way.

In preparation for the trial, Crowell & Moring, a law firm in Washington, D.C., that represents the Power Authority, hired eight scientists as expert witnesses. They submitted written testimony in June that no significant health effects would result from exposure to the Marcy South transmission line. According to *Microwave News,* none of these scientists had ever performed any research on the biological effects of 60-hertz electric or magnetic fields. In the July–August issue of the newsletter, Louis Slesin provided a synopsis of the conclusions that seven of the eight had drawn in their written testimony.

Dr. Stuart Aaronson, chief of the laboratory of cellular and molecular biology at the National Cancer Institute, in Bethesda, Maryland, reported that there was "no scientific basis for concluding that power-frequency electric and magnetic fields induce any consistent effects on cell growth properties in culture or in vivo that are associated with the acquisition of malignant properties." (In arriving at this opinion, Aaronson declared that Jerry Phillips's studies showing that cultured human cancer cells proliferate in the presence of such fields were seriously flawed.)

Dr. Richard Bockman, a researcher and attending physician at the Memorial Sloan-Kettering Cancer Center in New York City, stated that "no endocrine or metabolic disorders can be demonstrated in animals or humans exposed to EMFs at the levels generated by electric power lines." (His statement ignored studies conducted by Richard Phillips and his associates at Battelle, showing that ELF radiation could inhibit the pineal gland's secretion of melatonin, the hormone that regulates circadian rhythms and has been shown to inhibit tumor growth.)

Roswell Boutwell, a professor in the Department of Oncology at the University of Wisconsin, in Madison, concluded that "there is no scientific basis for the assertion that power frequency electric and/or magnetic fields are either cancer initiators or promoters." (This ran counter to the conclusions drawn by Wertheimer, Milham, and Savitz on the basis of their epidemiological studies, and by Adey, Jerry Phillips, and others on the basis of experimental investigations.)

Dr. Edmund A. Egan II, a professor of pediatrics and physiology at the University of Buffalo, concluded "with confidence" that "electric and magnetic fields associated with electric power generation, transmission and use have no ill effects on reproduction or prenatal development." (He found himself able to draw this conclusion despite the fact that the three-generation mouse study conducted by Andrew Marino had been partly confirmed by the three-generation swine study and other experiments conducted by Richard Phillips at Battelle.)

Dr. Lucius Sinks, a specialist in pediatric cancer at the National Cancer Institute, reported that "after consideration of genetics, hematology, immunology and epidemiology taken as a whole, it is my opinion that there is no causal relationship between exposure to power frequency electric and magnetic fields and childhood cancer." (Sinks arrived at this opinion despite the fact that three out of the four studies of childhood cancer published in the medical literature have shown an association between cancer in children and their exposure to alternating magnetic fields from nearby electrical distribution lines.)

Dr. Margaret Tucker, an epidemiologist at the National Cancer Institute, declared that recent research had yielded "no persuasive scientific evidence of increased risk of cancer in children or adults from exposure to power frequency electric and/or magnetic fields." (Tucker has also testified in behalf of the Potomac Electric Power Company in a hearing involving a 500,000-volt power line in Maryland; in behalf of the Mississippi Power Company in another lawsuit involving power-line radiation; and in behalf of utilities in Colorado and Arizona.)

Herbert S. Terrace, a professor in the Department of Psychology of Columbia University, reported that "Efforts to show that power frequency electric fields influence the performance of learning behavior have proved nil.'' (Professor Terrace's review failed to include the study that had been conducted for the New York Power Lines Project by Kurt Salzinger of Polytechnic University, in Brooklyn, who had found significant permanent changes in the learning ability and performance of rats exposed *in utero* and after birth to 60-hertz electric and magnetic fields. Terrace told *Microwave News* that he had read Salzinger's study but had ''theoretical and procedural'' problems with it.)

When the trial opened before Judge Peter A. McCabe, on September 8, 1988, Professor Chatkoff, of the University of Texas, had testified for the plaintiffs that a magnetic field of three milligauss or more in strength would be present at a distance of 500 feet or more from the Marcy South transmission line. At the trial, Andrew Marino testified that in his opinion no institutional review board in the United States would approve an experiment in which human beings were exposed to such electromagnetic fields if their informed consent had not first been obtained. Jerry Phillips testified about his experiments showing that cultured human cancer cells proliferate when exposed to power-line electric and magnetic fields. Dr. Harris Busch said that he believed it was ''dangerous'' to live in an area where the magnetic-field strength was three milligauss or more. The final witness for the landowner plaintiffs was Dr. Lennart Tomenius. He described his study linking childhood cancer in Sweden with exposure to electromagnetic fields from power lines, and also stated that he believed three milligauss to be a hazardous level of exposure.

Professor Terrace testified in behalf of the New York Power Authority that there was no reasonable scientific basis for people to fear transmission lines. When asked whether any of the studies he had reviewed had found negative effects in animals, he replied, ''None whatsoever.'' He went on to claim that ''There were some that suggest animals find these stimuli positive,'' and were ''attracted to them.'' Terrace's opinion that power-line radiation posed no health hazard was supported by Dr. Kenneth Zaner, an assistant professor of medicine at the Harvard Medical School. He testified that from a biophysical point of view, there was no evidence of a health risk from power-line electric and magnetic fields.

During the autumn, the Power Authority complied with Slesin's request that it furnish him with the exact amounts that it had paid to each of its expert witnesses, and Slesin published the tally in *Microwave News*. Dr. Aaronson, of the National Cancer Institute, received $70,250.98 for his testimony; Dr. Bockman, of the Memorial Sloan-Kettering Cancer Center, $56,950.70; Boutwell, of the University of Wisconsin, $74,256.74; Dr. Egan, of the University of Buffalo, $24,338.92; Dr. Sinks, of the National Cancer Institute, $41,083.42; Professor Terrace, of Columbia University, $54,153.69; Dr. Zaner, of the Harvard Medical School, $33,512.74; and Dr. Margaret Tucker, of the National Cancer Institute, $12,978.04 on the basis of preliminary billing.

The total amount of money paid to those eight witnesses was over $350,000. The plaintiffs paid their five witnesses a total of just over $60,000: Dr. Busch received $20,000; Chatkoff between $5,000 and $6,000; Marino $20,000; Phillips approximately $6,500; and Dr. Tomenius $9,000.

On December 29, 1988, Ross Adey wrote a letter to Dr. James B. Wyngaarden, the director of the National Institutes of Health (NIH) in Bethesda, Maryland, telling him that Drs. Aaronson, Sinks, and Tucker, of the National Cancer Institute, had testified at the trial, and how much they had been paid for doing so. (Adey also informed Wyngaarden that Dr. Mark Israel, chief of the National Cancer Institute's Molecular Genetics Section, had testified for a power company.) He then quoted from NIH regulations dealing with outside work and activities, which state that "Total compensation from consulting with profit making organizations, including industry and law firms, and from testimony in private litigation, is limited to $25,000 per year, with no more than $12,500 from any individual or law firm."

After reminding Wyngaarden that the National Cancer Institute was planning to study the association between childhood leukemia and electromagnetic fields, Adey suggested that the actions of some of its scientists in testifying "as hirelings of vested interests" had clearly called into question the National Cancer Institute's ability to conduct or oversee objective research into the problem. "Not only have these NIH employees demeaned their role as scientific experts with prime responsibilities as public servants, but their testimony on this issue of electromagnetic field interactions with biomolecular systems, including a possible relationship to cancer, reveals that, without exception, they know absolutely nothing about the underlying physical mechanisms, nor about the impressive body of scientific evidence relating these interactions to normal and abnormal growth, including a possible relationship to epigenetic carcinogenesis," Adey wrote. He went on to tell Wyngaarden that "no matter what prestigious position a scientist may hold within any restricted field of science, it is clearly irresponsible to posture as an authority in realms beyond the area of one's own specialty, particularly when, as in this case, there are so very few who have the requisite professional background in the biological, physical and engineering sciences."

Early in February of 1989, Adey received a reply from Wyngaarden, saying that he had initiated a "full review" of the circumstances Adey had cited in his letter. The review did not begin until late May, however. Meanwhile, no one has explained how the National Cancer Institute can participate in an objective assessment of childhood leukemia and power-line radiation when several of its leading researchers have testified under oath that there could not possibly be a link between the two.

Chapter

# PROTEST

COMMUNITY OPPOSITION TO POWER lines had been mounted in the autumn of 1976, when farmers and landowners held a large rally in Fort Covington, New York, to protest the construction of the 765,000-volt Massena-to-Marcy transmission line by the Power Authority of the State of New York. In 1980, farmers and landowners in Minnesota launched a similar campaign against a 400,000-volt direct-current power line that had been constructed in the central part of that state, claiming that an unusually high number of people who were living close to the line were suffering from headache, fatigue, stress, and nausea. In the winter of 1982, the Bonneville Power Administration ran into much the same kind of opposition when it proposed to run a pair of 500,000-volt power lines through western Montana. In the spring of 1983, there were protests in the Eastern Townships region of Quebec, when Hydro-Quebec—a utility owned by the province—announced that it intended to build a 450,000-volt direct-current transmission line across the area to the Vermont border, so that power generated by the huge $15 billion hydroelectric project that was nearing completion at James Bay, 650 miles north of Montreal, could be brought to New England. "We are being asked to sacrifice our countryside—and possibly our health—to power the air conditioners of Boston," Gary Caldwell, a sheep farmer in Ste. Edwige, told the Boston *Globe* at the time. Spokesmen for Hydro-Quebec denied that there was any health hazard, and other proponents of the power line, who included Premier René Levesque, pointed out that the debt-ridden province desperately needed the $5 billion in revenue that the line was expected to produce.

Five years later, the transmission line had become a fait accompli in the Eastern Townships, and been extended to the border of Maine. At this point, it began to exert a domino effect in the United States. In the summer of 1988, hundreds of residents of western Maine formed an organization called "No Thank Q Hydro-Quebec" to protest the proposal of the Central Maine Power Company, which had already signed a $4 billion twenty-nine-year contract to buy electricity from Hydro-Quebec, to construct a 145-mile-long, 450,000-volt direct-current transmission line to carry the power from the Canadian border to the southern part of the state. "I don't think western Maine and the people in this part of the state need to be guinea pigs," Kathy Sutton, a public health nurse from Roxbury, told *The New York Times*. "We're talking about families living next to this for a lifetime with children growing up there." Dr. Stephen Bien, a family practitioner in the town of Jay, was especially worried about the studies showing a connection between power-line radiation and childhood leukemia. "My basic feeling is that we stumbled blindly into a lot of environmental disasters in the past," he told the Boston *Globe,* adding that more research on the health effects of power lines should be undertaken before the high-voltage transmission line was built. Meanwhile, as real estate brokers were predicting that construction of the line would seriously depress land values in the area, officials of Central Maine Power, who had launched a multi-million-dollar campaign to convince lawmakers, state regulators, and the public to support the deal with Hydro-Quebec, were assuring property owners that land values might even rise.

In January 1989, the three-member Maine Public Utilities Commission voted two to one to reject the deal, on the grounds that it was not economically attractive for the state. Some observers believed that the vote merely marked a temporary setback for advocates of the power line. In any event, Hydro-Quebec wasn't hurting. At the beginning of 1988, Premier Robert Bourassa of

Quebec announced that the utility had agreed to supply 130 billion kilowatt-hours of electricity to the New York Power Authority between 1995 and 2015. The deal was expected to generate more than $13 billion for Hydro-Quebec, and it undoubtedly accounted for the Power Authority's decision to build the 345,000-volt Marcy South Line, as well as its willingness to pay more than $350,000 to eight scientists for their testimony that electromagnetic fields from the line would pose no health hazard.

In 1985, Ontario Hydro—a utility owned and operated by the province of Ontario—announced that it planned to run twin 500,000-volt transmission lines and a 230,000-volt line through the middle of a community called Bridlewood, in the city of Kanata, about ten miles west of Ottawa. The utility said that it would use an existing 230,000-volt right-of-way, in which there was already a 230,000-volt line, and that the two new lines would be strung on 16-story towers. A year or so later, after an article about the health effects of power lines appeared in the Ottawa *Citizen,* some Bridlewood mothers organized the Bridlewood Residents Hydro Line Committee, and soon involved the community in a major effort to have the proposed transmission lines rerouted. During 1986 and 1987 the committee raised more than $24,000, drew up a petition with more than a thousand names, and arranged to have more than three thousand letters of protest sent to David Peterson, the premier of Ontario. For its part, Ontario Hydro claimed that rerouting the lines would be expensive, and mounted a scare campaign that conjured up the specter of major blackouts and hospital power failures.

In July 1987, two days after the results of the New York Power Lines Project research program were released to the public, the Ontario Municipal Affairs Minister announced that the provincial cabinet had decided to allow Ontario Hydro to build the lines as planned; and in December, the utility began construction. At that point, the dispute moved into the provincial courts.

Earlier in the year, Canadian Television (CTV) of Toronto hired Michael A. Persinger, the psychologist at Laurentian University who had been a member of the New York State Power Lines Project's scientific advisory panel, and Stan Koren, an electronics engineering technician at the university, to measure the electromagnetic fields from the existing 230,000-volt power line in Bridlewood, and to estimate how much they might increase when the additional lines were installed. Persinger and Koren's findings were unsettling, to say the least. Koren measured magnetic fields of between 18.8 and 22.5 milligauss at some children's jungle bars in a playground within the power line right-of-way, and found magnetic-field levels of about 2 milligauss at the Bridlewood School, which is situated just outside the right-of-way. Persinger calculated that magnetic-field levels of between 5 and 100 milligauss would be likely inside the school after the proposed power lines were installed and activated.

Persinger reported that significant biological effects had been observed in both epidemiological and experimental studies from chronic exposure to such magnetic-field levels. He said that the brains of the offspring of pregnant rabbits that had been exposed to the predicted levels of the proposed Ontario Hydro lines "showed malformations within regions that control organized motor movements," and that other animals exposed to similar fields had demonstrated "long-term changes in behavioral productivity." He pointed out that the magnetic field could totally penetrate the human body, but could not be detected by the normal person. Exposures to the predicted magnetic-field strength of the proposed lines "have affected an entire complex of chemical reactions that are well correlated with mood, concentration and the 24-hour (circadian) rhythm," he warned. He added, "Although several days of exposure are required to elicit these changes, an exposure of only one to two hours can evoke weak but discernible changes in the electric activity within deep structures of the brain."

Since Persinger had been a member of the Power Lines Project's scientific advisory panel—the group that had found itself unable to draw any specific conclusions about the health hazards associated with power-line radiation—his assessment of the potential health threat in Bridlewood could not fail to raise the question of whether the power-line hazard was somehow more serious in Canada than in the United States, or whether the members of the scientific advisory panel had found themselves constrained to avoid reaching any conclusions that might alarm the American public.

# REASON NOT TO TRUST

PERHAPS THE MOST SUCCESSFUL COMMUNITY opposition ever mounted against a high-voltage transmission line was organized in the autumn of 1987 by Citizens Against Overhead Power Lines, Inc., a group of homeowners in the Highline area of South Seattle, in Washington State. They waged a highly effective campaign against a proposal by Seattle City Light to construct a pair of two-and-a-half-mile-long, 230,000-volt power lines on either side of State Route 509, between South 98th and South 136th Streets.

In May of 1986, City Light had recommended that the two power lines be constructed, and had issued an environmental impact statement to support the recommendation. The statement was a 203-page hodgepodge of estimates, comparisons, conclusions, and assurances, all of which appeared to have been thrown together in a manner that was designed to sow as much confusion as possible about the potential biological effects of the proposed transmission lines. Anyone who took the time and trouble to read those portions of City Light's impact statement that described the magnetic fields that had been measured and estimated by the Bellevue, Washington, engineering and consulting firm of Wilsey & Ham might well have been alarmed by the firm's assessment of field strengths given off by the ordinary electrical distribution lines that had been serving neighborhoods in South Seattle for many years. Over five hundred homes in the area were estimated to contain fields of more than five milligauss, and nearly two hundred of these homes were estimated to have fields of between ten and nineteen milligauss. The magnetic-field strengths on 14th Avenue South were unusually high, because a pair of high-current distribution feeder cables passed along the street. Wilsey & Ham had estimated that the lowest magnetic field strength would be created if the proposed transmission lines were installed along Route 509, and that estimate undoubtedly accounted for City Light's recommendation that they be placed there.

As for the health hazards associated with these magnetic fields, City Light's impact statement offered little comfort. It said that the possibility of a correlation between magnetic field exposure and cancer could not be dismissed, and it acknowledged that the magnetic fields that would be produced within 100 to 150 feet of the proposed lines would be "stronger than the fields implicated in the early epidemiological studies."

On July 29, 1987, the Seattle City Council met and voted seven to one to accept City Light's environmental impact statement and its recommendation to build the transmission lines along State Route 509. By that time, a small group of Highline residents led by Thomas Manley, a retired teamster who lives on South 136th Street, had begun to contact families living along Route 509 and to organize Citizens Against Overhead Power Lines, Inc. Only eighteen people attended the group's first meeting, in July, but by November one hundred and fifty families had become members. On December 1, Manley wrote to Governor Booth Gardner, pointing out that although City Light's own impact statement had revealed a possible connection between electromagnetic radiation from power lines and brain damage and childhood cancer, the utility had nonetheless taken the position that it was an "acceptable risk" to build high-voltage power lines within a few feet of residential dwellings. Manley asked the governor's help "to avoid harming children, and to save the City and State the huge expense of future litigation and potential liability."

By then, City Light had begun to mount a campaign to counter the appeals being made by Manley's group. The campaign relied in part on the discredited notion that household electrical appliances were more dangerous than power lines. On November 29,

Dennis Gray, the manager of systems engineering for the utility, told the Highline *Times* that people were exposed to electromagnetic fields whenever they used electrical appliances in their homes. Gray said that a 230,000-volt transmission line would create a magnetic field of 33 milligauss ten feet inside some property lines, whereas a toaster created a magnetic field of 100 milligauss and an electric mixer one of 8,000 milligauss. He did not point out that the magnetic fields from toasters and mixers fall off sharply with distance, and, because they are present only sporadically in homes, do not constitute a source of chronic exposure.

On January 4, 1988, Manley wrote to Mayor Royer, telling him that the residents of Highline "have reason not to trust City Light's objectivity or motives." Manley cited statements made at public meetings by representatives of the utility who had declared that radiation from the proposed power line would be no more harmful than that from toasters and refrigerators, and that the earth's background radiation was stronger than that generated by transmission lines.

By the end of the year, City Light was having second thoughts about its new project. On January 7, Randall W. Hardy, the superintendent of City Light, wrote a letter to Norman Rice of the City Council's energy committee, telling him that the utility, "after a December 1987 review of load and systems reliability data, is proposing a delay in the construction of the Highline substation and the connecting 230 kV transmission lines."

Two weeks later, Clyde L. Slemmer, a project development engineer for the Washington State Department of Transportation, wrote Hardy to tell him that the state would not approve City Light's request for a franchise to construct overhead transmission lines on Route 509 because of traffic safety considerations, obstructions of possible highway expansion by the poles, public opposition to the proposal, and concerns about long-term health effects. "Last, but not least, we are concerned about possible litigation and our required hold-harmless provision," Slemmer told Hardy. "We are not sure how the City of Seattle could hold us harmless from litigation on accident or health effect claims and who would pay the costs of litigation and judgments."

What Slemmer did not fully address, but what promises to become a major legal problem, is who can be held liable for claims involving cancer and other health problems that may be caused by the electric and magnetic fields that are given off by high-current wires in neighborhood distribution systems across the nation. In addition, what will be the extent of liability for future claims involving cancer and other illnesses that may result from exposure to the low-level fields emanating from PAVE PAWS and other radars, and from electric blankets, electric mattress pads, and electrically heated water beds? Moreover, what will be the liability of the nation's vast computer industry for claims arising from illnesses that may be caused by exposure to the low-level electric and magnetic fields that are known to be produced by video-display terminals, of which there are now some thirty million in use in the United States alone?

# Computer Terminals

# A UNANIMOUS DECISION

THE ATTEMPT OF THE UTILITIES INDUSTRY to play down the hazards of exposure to electric and magnetic fields from power lines has been abetted by a reluctance on the part of many people to acknowledge that their health could possibly be threatened by invisible emanations from something they regard as both pervasive and indispensable. Indeed, so dependent are we upon the benefits of electricity, and so accustomed have we become to the vast spiderweb of the electrical distribution system surrounding us, that we have accepted without question the necessity and ubiquity of its presence. This, in turn, has made it easy for us to embrace without reservation virtually all the hundreds of electrically powered devices that have been introduced into our homes, our workplaces, and our environment, and to make no distinction between them, other than to be aware of the different tasks they are designed to perform.

Ironically, it was officials of the Navy who first drew attention to the fact that different electrical devices give off different magnetic fields. These officials claimed that certain household appliances, such as hair dryers and food mixers, generated magnetic fields that were far stronger—and therefore more potentially hazardous to the health of a user—than the magnetic fields that were expected to emanate from Project Seafarer. What they failed to point out was that, with the exception of the electric blanket and the dial-face electric clock, the strength of the magnetic fields from household appliances falls off rapidly with distance, and that most appliances, including hair dryers and food mixers, are not sources of chronic exposure, because they are used sporadically. Similar claims by the utilities industry, which was anxious to exonerate power lines as sources of chronic exposure to magnetic fields, found ready acceptance by gullible government health officials and undiscerning scientists. As a result, the public has remained largely uninformed about the true nature of the hazard. Small wonder, then, that very few of the millions of people who use video display terminals (VDTs)—devices whose prodigious capacity for data entry and retrieval and for word processing has revolutionized office work—have any idea that they are being exposed to potentially hazardous 60-hertz magnetic fields emitted by their machines.

From the outset, the manufacturers of computer terminals have been disinclined to measure or to give out information about the magnetic fields that emanate from their products. Indeed, only recently have any of them acknowledged that such fields can be produced by VDTs. Their initial reluctance can undoubtedly be ascribed to the fact that during the 1970s most people considered low-level magnetic fields to be harmless. Their continued reticence, however, appears to be engendered by a belated concern about the laws of strict product liability that exist in almost every state. These laws clearly stipulate that the manufacturer of any product is responsible for testing the safety of that product, for being familiar with the medical and scientific literature relating to its safety, for knowing any harmful consequences that may result from its use, and for advising a potential consumer or user about any such consequences by affixing a warning label to the product. They also stipulate that anyone who manufactures and sells a product that is unreasonably dangerous to a user is liable for any physical harm it may cause. Far from issuing any warning about the possible health hazards of working with VDTs, however, the makers of these devices have denied for more than a decade not only that any health hazards exist but also that any could conceivably exist. Meanwhile, as the result of the wholesale embrace of computer technology by a largely unsuspecting public, some 30 million VDTs have made their way into offices, homes, and schools across the nation.

A typical word processor consists of a video display screen and a typewriter keyboard, both linked to a computer that can act as a repository for data. The display terminal operates on the same principle as a television set: the image on the screen is produced by an electron beam generated in a cathode ray tube (CRT)—an evacuated glass tube containing an electron gun, which produces a narrow electron beam. This beam is accelerated and directed toward the front of the tube by high voltage—between 15,000 and 25,000 volts—from a step-up transformer known as the flyback transformer. When the beam strikes the inner surface of the CRT face, or screen, it interacts with a phosphor coating to generate a spot of visible light, which glows in the form of a letter or a number.

Most VDTs display images by dividing the picture, or frame, into lines, called raster lines, which appear on the screen as the electron beam sweeps from left to right and from top to bottom. The movement of the electron beam across the face of the screen is controlled by a horizontal and a vertical deflection system, which consists of two sets of coils that are wound like a yoke around the neck of the CRT. The coils produce magnetic fields when electric current flows through them. Since magnetic fields can deflect electrons, it is the adjustment of current in the coils that controls the movement of the electron beam. A horizontal-deflection coil moves the beam from left to right; a vertical-deflection coil moves it from top to bottom. Each time the electron beam arrives at the right-hand side of the screen, a synchronization pulse causes the beam to "fly back" quickly to the left side of the screen. Since most VDTs produce sixty 262-line pictures every second, the electron beam travels back and forth across the screen more than 15,000 times a second—resulting in a typical horizontal-scan frequency of approximately 15,000 hertz (15 kHz), which is in the very-low-frequency (VLF) radio-frequency range. The typical vertical-scan frequency is 60 hertz.

When VDTs are operating, they emit X-rays from their cathode ray tubes, but so much of this radiation is absorbed by the glass of the CRT that it is not considered to pose a health hazard. The other radiation given off by VDTs includes ultraviolet, visible light, infrared, microwave, radio-wave, ELF, and static electric fields. Most of it consists of pulsed VLF electric and magnetic fields of between 15,000 and 20,000 hertz (15–20 kHz), and pulsed ELF electric and magnetic fields of 60 hertz. The pulsed VLF radiation is produced by the flyback transformer and the horizontal-deflection coil. The 60-hertz electric and magnetic fields are generated in two ways: 60-hertz fields originating in the 120-volt current that powers the VDT are emitted by the machine's power transformer (since these fields decay rapidly over distance, they can usually be measured only in the immediate vicinity of the transformer); and much stronger 60-hertz magnetic fields are produced by the CRT's vertical deflection coil, which governs the up-and-down movement of the electron beam on the VDT's phosphor-coated screen. These magnetic fields provide the dominating waveform given off by VDTs. This has only recently come to light, however, because VDT manufacturers have long chosen to ignore the fact that 60-hertz magnetic fields are generated by the vertical-deflection coil.

During the 1970s, VDTs were installed in thousands of offices in the United States, including many newspaper offices, where they quickly became popular. Not only did they afford reporters and editors ready access to stored information, thus speeding up the editing and writing process, they also produced copy that could be set in cold type immediately or held for future use. But there were problems. As early as 1975, Swedish scientists reported unusual numbers of VDT operators complaining of eye discomfort, and in 1977, a study conducted by researchers at the Swedish National Board of Occupation Safety and Health found that 85 percent of a sample group of Scandinavian Airlines System employees who worked with the machines were experiencing blurred vision or temporary nearsightedness.

In the United States, one of the first indications that VDTs might have adverse biological effects came in late 1976, when two *New York Times* copy editors, aged twenty-nine and thirty-five, who had been working with the machines for twelve months and four months, respectively, were diagnosed by their ophthalmologists as having developed opacities in the lenses of both eyes— the precursors of cataracts. Subsequently, each of the copy editors was examined by Dr. Milton M. Zaret, an associate professor of ophthalmology at the New York University-Bellevue Medical Center. During the early 1960s, Dr. Zaret had discovered that men working with radar in the armed services and the electronics industry were developing opacification and cataracts on the posterior capsule of the eye as a result of their exposure to relatively low-level microwave radiation. Now Zaret found that the older of the two copy editors was suffering from bilateral, incipient radiation-energy cataracts, and that the younger man had one immature and one incipient radiation-energy cataract. As a result, the Newspaper Guild of New York charged that the VDTs posed a threat to the health and safety of Guild members who used them, and the issue was submitted to arbitration.

In February 1977, Wordie H. Parr, chief of the Physical Agents Effects Branch of the National Institute for Occupational Safety and Health (NIOSH), and some co-workers measured the ultraviolet, visible-light, infrared, microwave, and radio-frequency radiation coming from several VDTs at the *Times,* including the two sets previously used by the copy editors who had developed cataracts. Parr and his associates reported that the electric- and magnetic-field strengths of the radio-frequency radiation that was being emitted were too weak to be detected by their instruments at a distance of ten centimeters. As it turned out, they were trying to measure the strength of the fields in terms of milliwatts per square centimeter, even though VLF and ELF fields cannot be accurately measured in this manner.

When questions about the instruments were raised, the arbitrator selected a private consulting firm to conduct additional measurements, and a panel of three ophthalmologists to advise him about the medical aspects of the case. Using instruments that were capable of detecting VLF radiation, but not ELF fields, engineers from the consulting firm found that the power levels of the radio-frequency radiation being emitted by VDTs at the *Times* were below those of the ten-milliwatt-per-square-centimeter guideline that had been recommended in 1966 by the United States of America Standards Institute to prevent injury caused by the heating of tissue. For their part, the three ophthalmologists told the arbitrator that they believed it unlikely such power levels could cause cataracts to develop. The arbitrator then handed down a decision stating the VDTs "do not pose any ocular radiant energy hazard," and that "the vast majority of all employees may work in safety upon the VDT machines at the *Times.*"

During the next four years, a succession of government agencies and private consulting firms continued to measure VDT radiation with inappropriate equipment at various locations around the nation, and to make pronouncements concerning the safety of VDTs that had little foundation in fact. In January of 1980, Parr and some colleagues from NIOSH conducted tests on 136 of some 530 VDTs that were being used by employees of the Oakland *Tribune,* the San Francisco *Chronicle and Examiner,* and Blue Shield of California, which has its headquarters in San Francisco. They measured VLF electric-field strengths as high as 1,400 volts per meter near the flyback transformers of some VDTs, and magnetic-field strengths of almost 9 milligauss near the flyback transformers of several units. In the report of their findings, however, they discounted these readings as anomalous, claiming that a current flow between the transformers and their measuring meter had interfered with the meter's electronic circuitry. "The flyback transformer can emit RF fields in the frequency range from fifteen to one-hundred-twenty-five kHz, but there is no occupational exposure standard for this frequency range, and these frequencies have not been shown to cause biological injury," they wrote.

What Parr and his colleagues were saying, in effect, was that because no studies had been conducted to show that VLF radiation could cause biological effects, the VLF electric and magnetic fields emanating from computer terminals could be discounted as a potential health hazard. Since the electric and magnetic fields emitted by VDTs were below the detection limits of their measuring equipment, they claimed that "the VDT does not present a radiation hazard to the employees working at or near a terminal." Moreover, they concluded that routine surveys of video display terminals were not warranted. Parr underscored NIOSH's lack of concern about VDTs in a 1980 interview published in the Bergen County (New Jersey) *Record.* "We don't particularly give a damn about them," he told the newspaper. "It's not our responsibility to go out and test VDTs. We just don't think there's a radiation problem." He added, "To be quite honest, nobody knows a damn thing about that low a frequency."

Since there were then some five million VDTs in the workplace, and approximately seven million men and women working on them, the exoneration of the machines as a health hazard by officials of an agency that had been created by the Congress for the express purpose of discovering and investigating occupational health problems was a significant development. As such, it had the effect of encouraging similarly unwarranted assessments on the part of other officials who had been assigned the task of evaluating the potential radiation hazard from VDTs. In 1979, an industrial hygiene engineer from the New York State Department of Health, who had seen the NIOSH report, told officials of *Newsday*—the largest daily on Long Island—that it was "perfectly safe" for someone to sit in front of a VDT for eight hours a day. Six months later, a health physicist from a private consulting firm, who measured high levels of radio-frequency radiation close to some VDTs at *Newsday*, also concluded that "it is perfectly safe to sit in front of the CRT for a normal working week." These assessments tended to reassure union officials who represented people working with VDTs that radiation from the devices was not likely to pose a significant health problem.

Meanwhile, Milton Zaret had seen through the charade that was unfolding. "As if the CRT problem were not bad enough, its resolution is being hampered further by many different but not unrelated attempts for a quick fix," he told colleagues attending the International Symposium on Electromagnetic Waves and Biology held near Paris in the early summer of 1980. "As gross equipment defects are found by crude testing, manufacturers, who previously denied there could be any problem, now pronounce these have been corrected by newly installed shielding, which didn't exist in the original model. NIOSH keeps citing its arbitrary, contrived standards for human exposure to nonionizing (both ultraviolet and hertzian) radiation as being safe, while there is no ratiocinate reason to believe this to be true, but, moreover, human pathology to affirm the opposite view. Labor and management, although it is in neither's long-term interest to do so, join each other in reassuring the workers with the palliative that there is little cause for concern. And epidemiologists are attempting to ascertain how to keep score of the aftermath. Meanwhile, tragically, nothing meaningful is being accomplished regarding prevention!"

Between 1978 and 1981, repeated assurances that VDTs were safe were provided by researchers at the Bell Telephone Laboratories, Inc., in Murray Hill, New Jersey, who conducted measurements of radiation emitted by the machines. In June 1978, Ronald C. Peterson, an electrical engineer at Bell Labs, measured the electric-field strength and power density of radiation being emitted by some of the VDTs in use at the labs, in order to determine whether it might be harmful. He reported that most of the energy he measured was low-frequency radiation associated with moving the electron beam; that its power density was far below the recommended ten-milliwatt guideline; and that there was no experimental or epidemiological evidence that these levels could have any detrimental effects on the health of personnel using these devices.

In April 1979, Peterson and Max M. Weiss, a physicist at Bell Labs, published an article in the *American Industrial Hygiene Association Journal* in which they described the findings of a survey they had conducted of electromagnetic radiation given off by eight representative VDTs being used in the Bell Telephone system. According to Peterson and Weiss, radio-frequency emissions from the machines were "well below those from commercial broadcast stations in the same frequency region," and were not associated with any health hazard.

In October 1981, Weiss and other scientists from Bell Laboratories presented papers on VDT hazards at a health conference in Nashville, Tennessee. In his presentation, Weiss said that extensive measurements of electromagnetic radiation from VDTs by researchers at Bell Laboratories, Duke University, and various federal agencies had led to the "unanimous conclusion" that VDTs "do not represent a health hazard from any radiation exposure caused by their use."

A day after Weiss spoke at the Tennessee health conference, a five-page memorandum about VDT radiation was circulated to employees of the Bell Telephone Laboratories. The memo was signed by B. L. Wattenbarger, an official at the labs, who informed the employees that scientists considered the radio-frequency radiation emitted by VDTs to be "harmless because it has no known detrimental effect on the human body." He said that people living in urban areas, especially near radio and television broadcast stations, "have been exposed to this type of radiation continuously for many years now, with no known harmful effects." The total radio-frequency radiation from a VDT "was slightly less than from the TV set," he maintained, and he assured them that VDT radiation would not turn out to be a hazard that would only be recognized after years of exposure, as in the case of asbestos. "The types of radiation that are emitted are weaker than other common sources of the same kinds of radiation, such as the sun and radio and TV broadcast signals," Wattenbarger's memo said. "Furthermore, there is nothing new to pose an unknown hazard—we have a great deal of knowledge and experience with all types of VDT radiation.''

Since there were then well over 100,000 VDTs in operation in the Bell Telephone system—many of them being used by directory assistance operators—it was perhaps understandable that Wattenbarger and his superiors should seek to reassure Bell employees that the company was looking out for their health and best interests. The fact remains, however, that for all of Bell's knowledge and experience, no Bell scientist said anything about the 60-hertz electric and magnetic fields that were being emitted by the machines, even though there were by then many articles in the medical literature—among them Nancy Wertheimer and Ed Leeper's 1979 study showing an increased incidence of cancer in children who had been exposed to low-level magnetic fields from high-current electrical distribution wires—to suggest that 60-hertz electromagnetic fields might be hazardous to health.

# 34

# THE EXPECTED
# UNEXPECTED

IN THE SUMMER OF 1980, concern about the possible adverse
health effects of VDTs took an ominous turn when it was learned
that within the previous year four out of seven pregnant VDT
operators in the classified advertising department of the Toronto
*Star* had given birth to infants with defects. One baby was born
with a clubfoot; another with a cleft palate; a third with an under-
developed eye; and the fourth was afflicted with multiple heart
abnormalities. All had been born between October and December
1979 to women ranging in age from the early twenties to mid-
thirties who said they had quit smoking and taken no drugs during
their pregnancies. During the same period, three employees at
the *Star*, who did not work on VDTs, gave birth to normal
babies.

As might be expected, an investigation of the unusual cluster
of congenital malformations was launched by officials of the Toronto Health Department at the request of the Southern Ontario
Newspaper Guild. At the same time, engineers of the Ontario
Ministry of Labor's radiation protection service proceeded to
measure radiation from the VDTs in the *Star*'s classified advertis-

ing department—there were about one hundred, most of which had been manufactured by IBM, of Armonk, New York—and also the radiation from nearly two hundred video display terminals that had been installed elsewhere at the newspaper.

As the investigation got under way, Dr. Bruce Dickerson, corporate director of health and safety for IBM, told an official of the Newspaper Guild that he considered the possibility of emissions from VDTs causing birth defects to be "totally new and without scientific predictability," and contrary to all the science developed to date. During the next few weeks, Canadian health officials seemed to fall over one another in their haste to absolve the machines of any blame. Gerald Caplan, coordinator of the Toronto Health Department's advocacy unit, issued a report stating that his group had thoroughly investigated the evidence concerning VDT hazards in the United States and Canada. "We left no stone unturned to find a case where a single machine emitted a dangerous level of radiation," Caplan declared, adding that "there is not a single scrap of evidence to indicate any danger from VDT radiation."

Anthony Muc, the chief of the Ontario Ministry of Labor's radiation protection service, who was in charge of measuring ionizing and non-ionizing radiation from the VDTs at the *Star*, also reported that none of them emitted hazardous levels. "Radiation from the VDTs cannot be the link in the four cases of birth defects discovered at the *Star*," he told a reporter from the newspaper. Saying that it was "time to draw the line," Muc recommended that the Labor Ministry stop rushing out to test VDTs whenever someone calls. Soon thereafter, he assured another reporter that "the body burden of radiation carried by one's bed partner exposes you to more radiation than working with a VDT all day." Muc's impatience with having to investigate the machines was echoed by Dr. P. J. Waight, acting director of the Canadian federal government's Radiation Protection Bureau, in Ottawa. "How many times do we have to go through this before people accept the evidence of studies in Canada, the United States and elsewhere?" Waight asked. "The machines are safe." He added, "We're getting sick of this kind of inquiry."

C. Eugene Moss, a health physicist at NIOSH, went even further. He not only agreed with Waight that the machines were safe, but also absolved them in a way that must have gladdened corporate hearts at IBM. He told the Toronto *Globe and Mail* that it was "absolutely impossible" for the IBM machines at the *Star* to cause birth defects.

During the next two years, seven unusual clusters of birth defects and miscarriages involving women who operated VDTs were reported in Canada and the United States. Between February 1979 and February 1981, seven out of thirteen pregnant women who worked part time at Air Canada's check-in counter at the Dorval Airport, in Montreal, miscarried. When this became known, Frank Stevens, Air Canada's manager of airport services, said, "We are convinced as a company and as a country that there is no radiation problem." His claim that all of Canada agreed with the company was apparently based on an article published in the *Canadian Medical Association Journal* by Dr. Ernest Letourneau, who had become the director of the government's Radiation Protection Bureau, which had surveyed more than two hundred VDTs. Letourneau wrote that "The best advice a physician can give a patient about VDTs is that they are no more dangerous than the monochrome television sets found in homes, and that they carry no radiation hazard." He suggested that the rapid proliferation of VDTs had led to "much unnecessary concern, particularly among women, that VDTs could be detrimental to the health of the operator and, in the case of a pregnant woman, her fetus." Still another federal health official who chose to exculpate the VDTs was Dr. Michael H. Repacholi, head of the Bureau's non-ionizing radiation section. "I cannot conceive of cataracts or birth defects coming from VDTs," he declared. (Six years later, he would be chairman of the World Health Organization task force that maintained it safe for men, women, and children in the general public to be exposed to power-line magnetic fields as high as 2,000 milligauss.)

During the summer of 1980, while the investigation of birth defects at the Toronto *Star* was still in progress, a consulting physician at Sears, Roebuck's southwest regional office in Dallas, called the Public Health Service's Centers for Disease Control (CDC) in Atlanta, to report that during the previous nine months six out of ten pregnant women who worked at the regional office had experienced spontaneous abortions, and that another employee had delivered a premature infant who subsequently died. Upon investigating, CDC officials determined that eight out of twelve pregnancies that had been conceived over a fourteen-month period by women working in Department 168 of the Sears office—a large room containing twenty-five VDTs in one corner —had ended either in miscarriage or in neonatal death. They also learned that during the first six working days of every month, almost all of the sixty-nine women who worked in Department 168 producing a monthly financial statement for the region spent at least six hours a day at the VDTs, and that for the remainder of the month, about a third of them continued to use the machines while the rest were assigned different tasks.

None of this, however, impressed the CDC officials as significant. "Even though we found a definite association between working in Department 168 and having an adverse reproductive outcome, we consider video display terminals an unlikely cause," they wrote in their report of the investigation. They went on to assert that "No previous clusters of spontaneous abortion due to VDT exposure have been reported," and they dismissed the birth defects that had occurred among the VDT operators at the Toronto *Star* by noting that all four of the abnormalities were of different types, and that two of them—cleft palate and clubfoot —were likely to be caused by agents other than radiation.

The CDC investigators chose not to measure radiation being emitted by the VDTs at Sears but to rely on the flawed NIOSH reports stating that radiation levels from VDTs were either below recommended guidelines or too weak to be detected. On this basis, they declared that "no causal link could be made between use of VDTs and subsequent spontaneous abortion." They concluded that "our findings may represent an 'expected-unexpected' cluster." (Just how unexpected can be gathered from the fact that since the expected rate of adverse pregnancies in the United States is 16 out of every 100, the estimated probability that 8 out of 12 pregnancies would be unsuccessful by chance is 6 in 10,000.) Dr. Nancy J. Binkin, an epidemiologist with the Abortion-Surveillance Branch of the CDC's Family Planning Evaluation Division, who had participated in the investigation, explained the CDC's conclusion in the following manner: "With 7 million VDT workers in the United States, many of whom are women of reproductive age, we would expect to see—on the basis of chance alone—several clusters similar to the one reported at Sears."

As it happened, Dr. Binkin's expectation had already become a reality. In December 1980, it was learned that between October 1979 and October 1980, three cases of congenital malformation and seven cases of first-trimester miscarriage had occurred among pregnant workers at the Marietta, Georgia, regional head-quarters of the Department of Defense's logistics agency. The cases of malformation included a child afflicted with hydrocephalus (enlargement of the head caused by the accretion of fluid in the cranium); a child born with a slightly asymmetrical head and an eye defect; and a child suffering from a defective pulmonary artery who died two months after birth. The 10 women whose pregnancies had ended adversely worked in a single building in which 206 women below the age of 45 were employed. According to a department safety and health manager, the amount of time each of the pregnant women had worked on VDTs each month ranged from a few minutes to full time.

Medical doctors from the Army's Environmental Hygiene Agency, who conducted an investigation of the Marietta cluster in February 1981, concluded that it was "an unusual statistical event for which no explanation can be found," and that it appeared "unlikely" to be related to a common work exposure. In their report, the Army doctors stated that the information gained from questionnaires filled out by the ten women with adverse pregnancy outcomes and eight co-workers who had experienced normal pregnancies "suggests that VDT exposure was not associated with adverse pregnancy outcome." Like the CDC doctors who investigated the miscarriage cluster at Sears, Roebuck, they failed to make any measurements of radiation emanating from the VDTs in question, relying instead on the flawed NIOSH reports that said there was no radiation hazard from VDTs.

By the end of 1982, Dr. Binkin's expectation that there would be more unexplained clusters of congenital abnormalities and miscarriages had come true several more times. Out of seven consecutive pregnancies occurring between May 1978 and October 1982 among clerks in the accounting department of the Surrey Memorial Hospital in Vancouver, British Columbia, there were three miscarriages, one premature birth, one infant born with a clubfoot and in need of eye surgery, and one baby suffering at birth from bronchitis. The seven women worked in the same room, and each of them had a VDT on her desk. At the request of the British Columbia Hospital Employees' Union, Hari

Sharma, a professor of chemistry at the University of Waterloo in Ontario, tested VDTs at the hospital, and found them to be emitting high levels of low-frequency radiation. Professor Sharma had previously recommended shielding the flyback transformers of VDTs and not allowing pregnant women to work at unshielded machines.

Between 1979 and 1982, seven pregnant women who worked on VDTs in the Office of the Solicitor General in Ottawa also experienced adverse pregnancy outcomes. Four of them miscarried, one gave birth prematurely, and two delivered infants who were suffering from respiratory disease. During that period, an eighth woman in the Solicitor General's Office, who did not use a VDT, gave birth to a healthy baby. However, Dr. Ian Marriott, a consultant for the Canadian government's Department of National Health and Welfare, ruled out VDTs as a possible cause of the problem, because, as he put it, the study of hazards from the machines "has been done so often by so many people around the world." A few months later, a task force appointed by the Canadian government had recommended that all VDT operators be given the option of performing alternative work during pregnancy.

Over those same three years, additional clusters of birth defects and miscarriages were reported in the United States. Chief among them was one at the Pacific Northwest Bell Telephone Company in Renton, Washington, where three out of five pregnancies occurring between July 1980 and August 1982 among women working with VDTs ended badly. One woman delivered a stillborn baby; a second woman gave birth to an infant suffering from spina bifida (open spine); and a third woman gave birth to a child afflicted with Down's Syndrome, a devastating defect known to occur as a result of exposure to X-rays, and also linked with exposure to microwave radiation.

When the Renton cluster was first discovered, officials of the Communications Workers of America tried unsuccessfully to negotiate with Pacific Northwest Bell and the Washington State Department of Health for a joint epidemiological study of the problem. (Earlier in 1981, officials of the Communications Workers of Canada and Bell Canada had worked out an agreement under which women would be allowed to stop working on VDTs during pregnancy.) Shortly after the cluster was reported, NIOSH officials announced that they were planning to conduct an epidemiological study of the impact of VDTs on pregnant women. Jay Bainbridge, assistant director of the Institute's Division of Surveillance, Hazard Evaluation, and Field Studies, said that while there was no evidence of any radiation hazard from VDTs, "We would like to resolve this issue once and for all."

# ANYONE WHO'S PARANOID

To SOME EXTENT, the reluctance of NIOSH and other governmental health agencies to thoroughly investigate the radiation hazard from VDTs can be traced to the outcome of a two-day congressional hearing on the potential health effects of VDTs and radio-frequency heaters and sealers—machines that use high-power radio-frequency radiation to melt and mold plastic and other materials in a wide variety of industrial processes. The hearing was held in May 1981 by the Subcommittee on Investigations and Oversight of the House Committee on Science and Technology. Testimony that workers who operate radio-frequency heaters and sealers were being routinely exposed to hazardous levels of radiation drew expressions of concern and outrage from Representative Albert Gore, Jr., the chairman of the subcommittee, and from several other committee members. By contrast, testimony concerning the radiation hazard from VDTs—about which there was considerable uncertainty and disagreement—drew a far more measured response, in spite of the fact that some of this testimony clearly indicated that governmental research into VDTs was seriously flawed.

One of the first witnesses to testify about the VDTs was John

C. Villforth, director of the FDA's Bureau of Radiological Health, who appeared before the subcommittee as a member of a panel that included Dr. Zaret. Villforth, who would later be quoted in *Science* as saying that "One of the beautiful things about radiation [is that] anyone who's paranoid can blame their troubles on it," told the subcommittee that a recent bureau survey showed that VDTs "do not pose a significant radiation hazard to persons who operate them." The survey to which he referred had found that 95 percent of the radio-frequency energy emitted from the VDTs was in the VLF range between 15,000 and 125,000 hertz. The bureau officials who carried it out had acknowledged in their report that no research had been conducted on the biological effects of VLF radiation, but they had nonetheless managed to conclude that VLF radiation "interacts only slightly with the human body, so that significant biological effects are unlikely." By way of explaining how they had reached this conclusion, Villforth told the subcommittee that radiation levels from VDTs that had been tested were "well below existing standards for the general public," with the exception of a few units that emitted X-rays in excess of the television receiver performance standard.

When Villforth went on to talk about the machines that emitted X-rays, Representative Gore broke in to ask if it was true that there were no standards for VLF emissions in the 15- to 120-kilohertz range.

"That is correct," Villforth replied, and added that the BRH had compared other emissions from VDTs, such as light, ultraviolet radiation, and ultrasound, with existing governmental and industrial standards.

Gore brought him back to the question at hand. "If there is no standard for the emissions in that range, then why is it reassuring to you that the nonexistent standard is not violated?" he asked.

"I think the answer is that we have a variety of knowledge from the scientific community that does not address the question of standards at all," Villforth replied. "The voluntary scientific community has not felt a need for standards in this area based on non-observance of biological effects."

"But you say yourself on page seventeen of your statement, 'Research on bioeffects for the fifteen to twenty-five kilohertz range is lacking.' "

"Yes," Villforth admitted.

"Okay. Now you say that it doesn't violate the standard, but there is no standard. You say that the reason there is no standard is because you have no reason to establish a standard, but then you say we have no reason to believe that it is safe either, in that frequency range, because we don't have any research. Now, it may be that your reassurance is valid, but it doesn't appear to be based on any scientific evidence at all. Would you agree with that statement?''

"Well, not completely," Villforth said. "Let me ask Dr. Andersen, who has had more experience in the biological effects area, to elaborate on this particular point, if I may."

"Maybe we had better come back to that, because I don't want to interrupt the panel," Gore said. "We can swear in another witness now, but let me go through the panel. Dr. Zaret, can I jump to you and ask you that question? How do you respond to that?''

"It is idiotic," Zaret replied.

"I can't hear you," Gore said.

"The statement is idiotic," Dr. Zaret said. "If you have no standard, you don't need a standard. We don't need it because there is no research to show that you need it. We haven't done the research. It makes no sense."

When Zaret's turn came to testify, he showed the subcommittee slide photographs of several cases of posterior capsular cataracts that he had diagnosed among VDT operators and air traffic controllers, who also work with cathode ray tube viewing screens. He then lashed out at the failure of governmental agencies to deal with the VDT radiation problem. "What is indefensible is the total failure of those agencies in our government having the responsibility to protect the general public as well as the worker," he said. "Organizations like BRH, OSHA, and NIOSH have been consistently wrong from the start, and their only apparent interest has been to obfuscate matters. Rather than having the good sense to keep quiet until they learned something, and actually trying to learn some things of real value, they have issued statements, reports, and proclamations that bring such discredit to themselves that they lose all credibility."

According to Zaret, the Bureau of Radiological Health's conclusion that significant biological effects were unlikely to occur in the VLF frequency range was "unwarranted." He emphasized the fact that the bureau had tested only a small number of VDTs, and that these machines had been tested one at a time and not in their usual setting. He went on to point out that in the real-life situation of the workplace, "many VDT operators could be irradiated from several surrounding VDTs at the same time they were being irradiated by the VDT in front of them."

Except for Zaret and some labor union officials who called for epidemiological studies of VDT operators, the thirty or so other experts from government and industry who testified at the hearings stated that VDTs posed no radiation health hazard. The representatives from the business community were adamant on this point. Vico Henriques, president of the Computer and Business Equipment Manufacturers Association, declared that "radiation from VDTs is within safe limits," and John Rankine, director of standards and data security for IBM, said that because of "rigid design testing, manufacturing checks, and company-sponsored research," he was certain that VDTs did not present a radiation health hazard.

At the conclusion of the hearings, Representative Gore summed up the attitude of the subcommittee members toward what they had heard about VDTs by saying, "The preponderance of evidence suggests that VDTs are safe, but we will encourage the Bureau of Radiological Health to set appropriate standards for VDTs and do more research on low frequency effects."

Just how vigorously Gore and his colleagues pursued the matter is open to question. Certainly, there is no evidence that he or any of his fellow committee members ever tried to encourage the Bureau of Radiological Health to set standards or conduct further research on the biological effects of the low-frequency radiation emitted by VDTs. As a result, while the number of VDT users in the United States increased from seven million to some thirty million over the next eight years, almost no studies were undertaken to fill the vacuum of knowledge about the health effects of the radiation known to be emanating from the machines.

Still another reason for the failure of NIOSH, OSHA, and the Bureau of Radiological Health to seriously investigate the potential radiation hazard from VDTs can be laid at the doorstep of the National Academy of Sciences. Shortly before the congressional hearing, NIOSH officials had asked the National Academy to review the existing studies of visual problems encountered by VDT workers and to suggest ways of resolving the question of whether the machines could damage the eyes. In response, the

Academy's National Research Council appointed a twelve-member panel of experts (only three of whom were ophthalmologists) to assess the impact of VDT use on the vision of workers. They issued a 273-page report in which they declared that their conclusions were severely limited by the small number of studies (seven) that the members of the panel had been called upon to assess, and because many of these studies had "substantial shortcomings in methodology."

This caveat notwithstanding, the members of the panel concluded that there was "no scientifically valid evidence that occupational use of VDTs is associated with increased risk of ocular disease or abnormalities, including cataracts," or that "the use of VDTs per se causes harm, in the sense of anatomical or physiological damage, to the visual system." (In reaching this conclusion, they dismissed the findings of Dr. Zaret, whom they did not bother to contact.) They went on to say that the levels of radiation emitted by VDTs were "highly unlikely to be hazardous," and they echoed the NIOSH conclusion that "routine radiation surveys of VDTs in the workplace are not warranted." They declared that "competent studies have found that the levels of radiation emitted by VDTs are far below current U.S. occupational radiation standards, and are generally much lower than the ambient radiation emitted by natural and man-made sources to which people are continuously exposed." To explain why no standards had been set for radio-frequency radiation in the VLF range that was known to be leaking from the flyback transformers of VDTs, they suggested that "For those forms of radiation for which few guidelines exist, there is generally little demand for standards, either because few people are exposed to such radiation or because there is no general concern that such radiation is hazardous at present levels of occupational or environmental exposure." Finally, they took refuge in the line used so often by the power industry, claiming that radiation levels from VDTs are "far below those emitted by many common electronic products or those present from natural sources in the environment."

As with most National Academy of Sciences studies, the findings of the panel received considerable coverage—especially in the nation's newspapers, where VDTs were by then much in use —and they have been widely accepted by the members of the nation's medical and scientific community. However, Charles A. Perlik, Jr., president of the Newspaper Guild, called the VDT report "misleading," and David LeGrande of the Communications Workers of America said it was "a real cop-out." A highly critical dissent has also been registered by Karl U. Smith, professor emeritus of psychology at the University of Wisconsin, in Madison, who has insisted that the Academy's report was seriously deficient and should not have been issued until the ocular problems of VDT operators had been studied in a scientifically acceptable manner. Professor Smith, as it happens, was the first scientist to investigate visual problems resulting from video viewing. He did so during World War II, when he was the associate director of an extensive investigation of cumulative perceptual fatigue among search-radar operators, who, like present-day VDT operators and air traffic controllers, worked at cathode ray tube viewing screens. Indeed, according to Smith, the only real difference between the tasks of the search-radar operators he studied forty-five years ago and the VDT operators of today is that the former worked at six-inch CRT scopes, on which blips representing target airplanes appeared as elevations of horizontal-scan lines, while VDT workers read words and figures on slightly larger VDT screens.

The perceptual fatigue study was a research project of the National Defense Research Committee—a branch of the Office of Scientific Research and Development that directed all military, industrial, and academic research connected with the war effort —and it was carried out under a contract with Yale University in a highly secret radar laboratory at Camp Murphy, near Hobe Sound, in Florida. It was undertaken to investigate widespread complaints by Air Force and Navy radar operators, who reported that they were experiencing blurred vision, irritated eyes, headaches, insomnia, irritability, impotence, and sterility. Smith and some colleagues at Camp Murphy designed a controlled experiment to measure and record the individual accuracy of search-radar operators in detecting and reading simulated aircraft blips that were displayed on their CRT scopes. The experiment, which became known as the Camp Murphy Project, demonstrated the effects of cumulative fatigue upon the performance of search-radar operators. Its findings, compiled in a secret 1944 study entitled "Radar Operator Fatigue: The Effects of Length and Repetition of Operating Periods on Efficiency of Performance," led the Navy and Air Force to establish and maintain, whenever possible, four-hour duty and four-hour rest schedules for all search-radar operators during the remainder of the war.

"Before undertaking the perceptual fatigue experiment, I personally carried out extensive field interviews with search-radar operators stationed at Camp Murphy, Clearwater, Tarpon Springs, Fort Lauderdale, Boca Raton, Fort Myers, and Orlando," Professor Smith said recently. "Virtually every one of these men complained of eye fatigue, eye discomfort, irritation of the eyes, insomnia, and irritability. Although radiation measurements were not taken, it was my opinion then, as it is now, that radiation given off by the cathode ray tube scopes and the radar transmitters, as well as the sixty-hertz electric and magnetic fields emanating from the large transformers that supplied power for the transmitters, were probably a factor in some of their complaints. It was also my judgment that their eye fatigue was caused by long duty periods of eight hours or more, and by a combination of stress-inducing factors. These factors included disturbed eye-and-head movement coordinations while engaged in a poorly designed and monotonous reading task; the necessity of sustained tunnel vision in a darkened room; restrictions on social life and activity; and a fear of the equipment with which they were working. It was clear to me that the fatigue and irritability of the search operators carried over into their domestic and sexual lives, and that these induced social disturbances probably exacerbated their visual complaints and were sufficient to produce headaches, impotence, and other psychophysiological and emotional disorders, including chronic diseases."

Professor Smith explained that human beings have evolved a muscular and visual system of reading books and other printed material that is not comparable to the demands of reading a vertical screen with enlarged letters and numbers. "The process of reading demands a continuous coordination of head movements and minute saccadic, or jerky, movements of the eyes, which occur in the span of milliseconds," he said. "The size of print has evolved gradually to permit sustained non-fatiguing movements of this sort. However, these tiny movements and their coordinations are thrown off completely when someone is working on a video display terminal. To begin with, enlarged saccadic movements of the eye are required to scan the printed material. Secondly, enlarged head movements are needed to read successive lines. Third, the VDT reading task demands a vigilant and tiring upright position of the head. Fourth, the reading distance is significantly increased. And, finally, the often blurred display of a VDT tends to keep the accommodation muscles of the eyes dancing in an attempt to focus upon the printed material and to fix its exact location in space. The eye operates as one of the most highly coordinated and integrated motor sensory systems of the body, and when its muscles become overly fatigued as a result of excessive VDT viewing, the blood circulation and the organic metabolism of the eye can be disturbed.

"For all these reasons, it is clear today, as it was forty-five years ago, that people engaged in cathode ray tube reading jobs should not be required to work at the machines longer than three or four hours at a stretch. Instead, office workers and air traffic controllers have been chained to their VDTs and CRT terminals, with the result that reports of visual complaints, behavioral problems, and chronic diseases among them have steadily increased."

Professor Smith finds it ironic that the Camp Murphy Project of 1943 remains the only comprehensive and sustained investigation of perceptual fatigue that has ever been conducted in the United States. "Why didn't the ophthalmologists, psychologists, and other members of the National Academy's vision panel consider this investigation in their review of the studies of visual problems encountered by VDT workers?" he asks. "Why haven't researchers at NIOSH and in the electronics industry used its findings as a basis for further exploring the cumulative effects of perceptual fatigue—a major potential public health hazard that is being created by the proliferation of VDTs in the workplace?"

# AN OCEAN OF DENIAL

WHEN MAX WEISS, OF THE Bell Telephone Laboratories, pro-
claimed that researchers in industry, government, and academe
had reached the unanimous conclusion that there was no health
hazard from VDT radiation, he was not exaggerating. By the
autumn of 1981, the presumption of benignity had been conferred
upon electromagnetic emissions from the machines by the arbitrator of a workmen's compensation proceeding; by the director of
NIOSH's Division of Biomedical and Behavioral Sciences; by the
chiefs of the Institute's Physical Agents Effects Branch and its
Radiation Section; by the head of the FDA's Bureau of Radiological Health; by scientists at the Public Health Service's Centers for Disease Control; by officials of OSHA and EPA; by the
director of the Radiation Protection Bureau of Canada's Department of National Health and Welfare; by members of the Ontario
Ministry of Labor and the Toronto Health Department; by hy-
gienists and scientists at Harvard University, Duke University,
the New York University Medical Center, and other leading academic institutions; by medical doctors from the U.S. Army's Environmental Hygiene Agency; by the medical directors of Air

Canada and IBM; by the president of the Computer and Business Equipment Manufacturers Association; by representatives of virtually every VDT-manufacturing corporation in the nation; by an official of the American Newspaper Publishers Association; by officers of *The New York Times*, the Toronto *Globe and Mail*, and other major newspapers; and, finally, by the members of the National Academy of Sciences review panel. Indeed, except for the outspoken Milton Zaret, the vote of confidence that had been given to the VDTs and to the low-frequency radiation emanating from them was virtually monolithic, and the sea of disbelief in which Nancy Wertheimer had felt herself cast adrift after she published her study on power-line magnetic fields and childhood cancer had become an ocean of denial.

In October 1982, however, a scientist in Canada came forward to take issue with the prevailing viewpoint and offer a warning about the presumed safety of VDTs. He was a biophysicist named Karel Marha, and his decision to speak out was particularly courageous in light of the fact that he had only recently arrived in Canada as an émigré from Czechoslovakia, where he had for twenty years been the director of the Department of High Frequency at the Institute of Industrial Hygiene and Occupational Diseases, in Prague.

Marha knew that his views on VDT radiation would be considered controversial. Back in 1969, he had visited the United States to deliver a paper at a three-day Symposium on the Biological Effects and Health Implications of Microwave Radiation, sponsored by the Medical College of Virginia, and held in Richmond. At that time, he told skeptical American researchers that he and his colleagues in Prague had found a wide variety of neurological problems, such as headache, eye pain, excessive fatigue, forgetfulness, and irritability, among people working in radar centers, in radio and television stations, and in factories where microwave-emitting devices were manufactured. According to Marha, these effects could be induced at power densities as low as one tenth of a milliwatt per square centimeter—one-hundredth of the recommended American guideline for occupational exposure. Because the effects of repeated irradiation were thought to be cumulative, and because large variations had been found in the sensitivities of different people, the Czechoslovak standard for exposure to microwaves incorporated a safety factor of ten, and was thus similar to the ten-microwatt standard that was in force in the Soviet Union, which was one thousandth of the American standard. According to Marha, pregnant women in Czechoslovakia were specifically prohibited from working in areas where these levels were exceeded.

Marha had described the reason for this prohibition at some length in a book entitled *Electromagnetic Fields and the Life Environment*, written with two of his colleagues and subsequently translated into English and published in 1971 by the San Francisco Press. In a section of the book dealing with the biological effects of radio-frequency energy, Marha and his co-authors pointed out that morphological alterations caused by microwave radiation could result in "changes of the reproductive cycle, in a decrease in the number of offspring, in the sterility of the offspring, or in an increase in the number of females born." After stating that irradiation of pregnant women and female animals apparently increases the percentages of miscarriages, they described a case of birth defect that occurred after a mother was treated with short wave diathermy at the beginning of her pregnancy. They then noted that "other authors have also reported that the [radio-frequency] field definitely impairs embryogenesis in both humans and animals, particularly in the beginning stages," and that the "development of the fetus is retarded, congenital defects appear, and the life expectancy of the infant is reduced."

After delivering his paper at the 1969 Richmond symposium, Marha was subjected to some tough questioning by his listeners. Among his interrogators was Lawrence Sher, of the University of Pennsylvania, who, unknown to most of the scientists attending the conference, was then serving on a secret advisory panel set up under the auspices of Project Pandora to consider the biological effects of low-level microwave radiation on Foreign Service personnel and other people who were living and working at the American Embassy in Moscow. Sher told Marha that "we are subjected to many possible insults from our environment," and asked him how he could possibly justify limiting microwave radiation exposure to intensities that precluded the smallest possible effect by a safety factor of ten. Marha replied in a fashion that epitomized the profound differences in thinking and approach to the problem between scientists in the countries of Eastern Europe and those in the United States. "Our standard is not only to prevent damage but to avoid discomfort in people," he said.

After coming to Canada in 1981, Marha worked for a year for Ontario Hydro, where he began to study radiation given off by video display terminals. In the summer of 1982, he left the utility to work at the Canadian Centre for Occupational Health and Safety—a non-profit information service in Hamilton, Ontario, which is financed by the federal government of Canada and run by representatives of industry, organized labor, and the federal, provincial, and territorial governments. On October 6, 1982, he delivered a paper entitled "The State of Knowledge Concerning Radiation Emissions from Video Display Terminals" at Carleton University in Ottawa, in which he disclosed information about the machines that had never before been discussed publicly by any scientist, health official, or VDT manufacturer in either the United States or Canada.

A soft-spoken, distinguished-looking man then in his early fifties, Marha told his listeners that the primary source of electromagnetic radiation from the VDT was (as had previously been reported) the machine's horizontal-deflection system and its high-voltage circuit, which generate voltage of sufficient magnitude to accelerate electrons in the electron gun of the cathode ray tube, and thus project an electron beam onto the VDT's phosphor screen. In order to achieve the correct beam deflection, he explained, the high voltage must be pulsed on and off at a rate of approximately 15,000 to 20,000 times a second, and this pulsating high voltage created electric and magnetic fields whose amplitudes were highest in the low-frequency and very-low-frequency bands. Marha pointed out that very few measurements with proper instruments had been made of the low-frequency radiation emanating from VDTs, but that a study in the United States had found electric fields within twelve inches of a VDT to be more than five times as strong as those allowed for occupational exposure in the Soviet Union, Czechoslovakia, Bulgaria, and the German Democratic Republic. He then revealed for the first time that, in addition to VLF radiation, VDTs also emit significant ELF radiation in the form of 60-hertz electric and magnetic fields from the vertical-deflection coil, and he described some recent measurements of 60-hertz magnetic fields from three different VDT sets, which had been made by Maria Stuchly and two colleagues from the Radiation Protection Bureau, in Ottawa. They had found 60-hertz magnetic fields greater than two milligauss at a distance of twelve inches from the viewing screens of two of the models; a field of just over one milligauss at the same distance from the screen of a third; and fields of approximately one milligauss twenty inches from the viewing screens of all three VDT units.

When Stuchly and her associates issued a report of their measurements, they said that when they had taken a reading close to the power transformer at the back of one VDT, they measured a magnetic field of more than 40 milligauss. Apparently, they were either unfamiliar with Wertheimer's findings, or chose to ignore them, because they went on to declare that ELF magnetic fields in the vicinity of VDTs "are of such low intensities that they are very unlikely to have any biological effect, let alone represent a health hazard." By way of justifying this conclusion, they said that VDT magnetic fields were comparable to those of other electrical devices, pointing out that a hand mixer had been found to emit localized levels of 92 milligauss. In this way, they managed to ignore the fact that VDT operators who often sit within a foot of their screens for hour after hour each day are being chronically exposed to magnetic fields of about the same intensity as those Wertheimer and Leeper found unduly often in the homes of children who had died of cancer in Greater Denver.

Marha went on to discuss an experiment that he and two Czechoslovakian colleagues had conducted twenty years earlier, which showed that pulsed electromagnetic fields, such as those emanating from radar transmitters and VDTs, were far more biologically potent than continuous-wave fields, such as those given off by radio and television transmitters. He then pointed out that there had been studies showing that low-frequency modulation of the pulsed electromagnetic field was of primary importance in the production of biological effects, whereas the high-frequency-carrier wave played only a secondary role. At the end of his talk, Marha drew some disturbing conclusions which effectively contradicted the rosy picture painted of VDTs by the vast majority of his fellow scientists. "Firstly, the VDT may produce electric, magnetic, and electromagnetic fields in nearly the entire non-ionizing band," he said. "The highest intensity fields can be found in the low-frequency spectrum. Most of these are pulsed fields or extremely low-frequency modulated fields. Secondly, all these fields are known to produce some biological effects. The complex evaluation of all possible factors, including combinations of different effects and all other hygienic and ergonomic factors near VDTs has not been done.''

Marha's speech at Carleton University was given almost no coverage in the Canadian press. Three months later, however, the Canadian Centre for Occupational Health and Safety issued a three-page news release about the VDT radiation hazard. The release stated that Marha and three other scientists at the Centre —Gordon Atherly, its president and chief executive officer; James T. Purdham, director of technical services; and Barry Spinner, manager of safety services—had called attention to the fact that some VDTs produced pulse-modulated VLF electromagnetic fields whose strength ''sometimes exceeds exposure limits set for other parts of the electromagnetic spectrum,'' and that there was scientific evidence to suggest that pulsed fields could be more harmful than non-pulsed fields. The release further warned that although there was no demonstrable link between VDT exposure and adverse pregnancy outcomes, Marha and his colleagues believed that governmental regulatory agencies should begin discussions with VDT manufacturers on the feasibility of shielding all sources of pulsed electromagnetic fields in new and existing VDTs, and that an interim standard of operator exposure should be established to require that electric fields at a distance of thirty centimeters (about one foot) from the nearest surface of a VDT not exceed 60 volts per meter. The Centre urged that ''These actions should be initiated, without waiting for the results of the further research, which, in any case, should be completed as quickly as possible.''

Incredible as it may seem, the Centre's news release was almost totally ignored by the media in Canada and the United States. As a result, the unprecedented warning and recommendations of Marha and his colleagues went virtually unreported in newspapers and on television newscasts in both nations. In an attempt to break through the wall of silence, Louis Slesin wrote a 600-word column about the Centre's white paper in the January–February 1983 issue of *Microwave News;* like the news release, it was disregarded. In his column Slesin revealed that the Centre's recommendation to limit VDT radiation exposure had drawn at least one reaction. It had come from the Canadian Business Equipment Manufacturers Association, in Toronto. On the day before the white paper was released, James Flood, the association's general manager, had sent the Centre a telex warning that its statement would create needless concern among VDT users, and asking that it be withheld. Flood insisted that "the Centre has proposed solutions to problems that don't exist," and that "there is no evidence indicating that VLF radiation is a hazard."

# 37

---

# BENIGN NEGLECT

DURING 1983, the Canadian Centre for Occupational Health and Safety published three more reports about the hazard of VLF radiation from VDTs. In the first of these, issued in April, Marha, Purdham, and Spinner pointed out that pulsed electric and magnetic fields emanating from VDTs were highly directional, the direction depending upon the design of the machine, and that for this reason, "it is evident that the exposure of other people near the VDT can be higher than the exposure of the operator." They went on to say that radiation measurements should be made not just in front of the viewing screen, as almost all measurements had been made until then, but also at distances of from 10 to 30 centimeters from all surfaces of the machines. After noting that radar waves modulated to pulse at ELF frequencies had been shown by Czech researchers to cause birth defects in mice, and by Swedish scientists to damage the cultured blood cells of humans, they observed that there had been "no systematic attempt to study the possible biological effects of exposure to VDT radiation."

In a second report, issued in July, Marha described ways in which the electric fields given off by VDTs could be effectively shielded. He suggested, however, that a simpler and more fool-proof way of preventing harmful irradiation by VDTs would be to "design the workplace in such a way that no one can sit or stand close to the side or behind the VDT unit." According to Marha, a distance of one to one and a half meters would be sufficient to afford protection. He also recommended that VDTs not be located back to front in a workplace "since individual operators could be exposed to emissions from nearby terminals," and that pregnant workers who had been relieved of VDT work "not sit close to the side or rear of a terminal."

In a third report, issued in December, Marha and David Charron, an electrical engineer at the Centre, warned that lead aprons, which were then being considered as protective gear for VDT operators, were totally ineffective in deflecting or absorbing VDT radiation.

As in the case of the news release that the Centre had issued at the beginning of the year, the three additional reports were almost universally ignored by the press in Canada and the United States. By now, it was plain that publishers and editors of major news-papers in both nations were avoiding stories about VDT radiation. Indeed, this state of affairs had been exposed in an article entitled "VDTs: The Overlooked Story Right in the Newsroom," which had appeared in the January–February 1981 issue of the *Columbia Journalism Review.* The article was written by Jeff Sorenson, director of the news syndicate for *The Nation,* and by Jon Swan, senior editor of the *Journalism Review,* and its subtitle left little doubt about their conclusions. It said, "If you want to learn about a health and safety controversy affecting thousands of journalists—and millions of other U.S. workers—don't rush out and buy a major daily."

Sorenson and Swan began by disclosing that the use of VDTs by the classified-advertising and editorial departments of the nation's newspapers had increased by about 50 percent annually since the devices had been introduced into the marketplace in the late 1960s. After reviewing the 1977 arbitration decision that VDT radiation was not a factor in causing cataracts to develop in the eyes of two copy editors at *The New York Times,* Sorenson and Swan took note of the fact that NIOSH had conducted tests on VDTs at the *Times* using instruments that were inadequate for measuring VLF and ELF magnetic fields. They went on to say that neither the *Times* nor *The Washington Post,* nor the Los

Angeles *Times,* nor the Chicago *Tribune,* nor the Chicago *Sun-Times,* nor the Baltimore *Sun,* nor, for that matter, any other major daily in their survey had reported on the arbitration hearings or on the fact that the NIOSH measurements were flawed.

Sorenson and Swan also pointed out that in October 1980, *The New York Times* had published a lengthy article entitled "Benign Radiation Increasingly Cited as Dangerous," which described television and radio broadcasts, telephone relay equipment, radar, and electric power transmission lines as sources of low-level non-ionizing radiation, but managed to avoid any mention of VDTs. They disclosed that approximately twenty newspapers in the nation—among them *The Washington Post,* the *Cleveland Plain Dealer,* and the Baltimore *Sun—were* paying for eye examinations for their employees. This, however, is what they had to say about how the *Sun* was covering the VDT issue:

> Like the *Sun-Times,* the Baltimore *Sun* has run only a single article on VDTs, a five-inch wire-service story that appeared last June. Given the fact that two *Sun* journalists have been diagnosed as having cataracts, this seemed very short shrift. We called the *Sun's* medical writer, Mary Knudson, to ask if she knew of any further coverage of the subject in her paper. "To my knowledge," she said, "that's all we've done, and I think that's wrong, because the *Sun* newsroom is aware that VDTs are controversial and especially because we at the *Sun* are about to become involved in a federal study by NIOSH to see if there is a higher rate of cataracts among VDT users than among non-VDT users." Knudson went on to say that last February she had offered to write a comprehensive account of the VDT controversy, but her editor told her this would be a conflict of interest since Knudson is chairperson of the health and new technology committee of the Sun Paper's Guild. "He said he would assign it to another reporter," Knudson said, "and that's the last I heard about it. The story hasn't been written."

Although highly critical of the way most newspapers were handling the VDT story, Sorenson and Swan gave credit for thorough coverage of the controversy to the Bergen County *Record,* which had run a comprehensive piece by Elliott Pinsley in March 1980. Pinsley wrote that the issue of VDT radiation "is clouded by emotionalism, a shaky federal standard, a lack of authoritative research on low level RF emissions, a lack of faith in the radiation monitoring devices currently in use, and a varying degree of public concern and candor among manufacturers." Sorenson and Swan also had praise for an article entitled "A Worrying Case of the VDTs," which appeared in the Canadian newsweekly *Maclean's,* in July 1980. It was written by Larry Black, a labor reporter for the Canadian Press, Canada's leading wire service agency, who quoted an airline union organizer as predicting that "cathode-tube operators are going to be the asbestos workers of the future."

In conclusion, Sorenson and Swan asked, "What has happened here?" They answered:

> The nation's newspapers are involved in a technological revolution; so, too, of course, are many other and much larger industries. This revolution is affecting the lives—and possibly adversely affecting the health—of thousands of journalists and millions of nonjournalists. At least some reporters are eager to cover various aspects of this large event, presumably in the same professional manner in which they would cover any other story. Meanwhile, there is a dearth of coverage. In some cases this may be because reporters genuinely do not see a story in VDTs. But it would not seem too farfetched to conclude that at the management level another factor is at work—that in their concern to protect a vast investment, editors and publishers are subordinating the public interest to a private one.

During the next few years, the *Columbia Journalism Review* continued to peer over the shoulder of the press and report on how the VDT story was, or was not, being covered. In its November–December 1984 issue, the *Review* published two articles and an editorial about the VDT health issue. One of the articles was written by Louis Slesin, the publisher and editor of the newsletter *Microwave News,* who pointed out that NIOSH had failed to initiate a study of pregnancy risks among VDT operators which it had promised to undertake two years earlier, and that NIOSH staff members had admitted in the spring of 1983—six years after NIOSH researchers had made their inadequate measurements at *The New York Times*—that they still had no ability to measure VLF radiation from video display terminals at job sites.

The second article was entitled "VDT Regulation: The Publishers Counterattack," and was written by Loren Stein, an associate at the Center for Investigative Reporting, in San Francisco, and Dianna Hembree, director of the Center's "Women in the 80s" project. Stein and Hembree disclosed that the American Newspaper Publishers Association, an organization representing some 1,400 newspapers in the United States—roughly 90 percent of the nation's daily and Sunday circulation—had joined the Computer and Business Equipment Manufacturers Association to form the Coalition for Workplace Technology, a powerful lobbying group, and that the Coalition was opposing measures to regulate VDTs then being introduced in various state legislatures. Many of the proposed safety bills would require employers to install anti-glare screens and metal shielding to protect VDT operators against radiation; to pay for mandatory eye examinations; and to transfer pregnant VDT operators, upon request, with no loss in pay, benefits, or seniority, to work that did not involve the use of VDTs. Stein and Hembree went on to say that George Cashau, the director of research for the Newspaper Publishers Association, had testified at a congressional hearing held by the Subcommittee on Health and Safety of the House Committee on Education and Labor in June 1984 that "inaccurate information" was responsible for employee fears about VDTs, and that Dr. Howard Brown, the medical director of *The New York Times,* had appeared before the subcommittee on behalf of the Association and testified that he was aware of "no medical evidence" of serious VDT-related health effects. They then disclosed that a computer search of the *Times, The Washington Post,* the *Christian Science Monitor, The Wall Street Journal,* the Los Angeles *Times,* and the Associated Press and United Press wire services revealed that none of them had published a single word about the warning of Marha and his colleagues at the Canadian Centre that the relatively strong pulsed VLF fields being emitted by the flyback transformers of some VDTs might have serious biological effects, or about their recommendation that metal shielding be installed on all VDTs.

Stein and Hembree found that a number of newspaper editorials had opposed the passage of VDT bills without mentioning that the publishers of these newspapers had a financial stake in the outcome. A typical editorial appeared in the Columbus (Ohio) *Dispatch* on February 13, 1984. Entitled "Assault on VDTs," it addressed itself to a VDT safety bill then before the Ohio State Legislature that would have required free eye examinations, rest breaks, and the transfer of pregnant employees from VDT work. "It's a good thing that some of the present members of the Ohio General Assembly weren't around when the pencil was invented,'' the editorial began. ''Had they been, the world's most basic writing instrument might today come equipped with a tip safety shield with see-through visor, work gloves, protective eyeglasses, ear shields, and an anti-chew guard.''

The editorial in the November–December issue of the *Columbia Journalism Review* took another view of the matter. Its authors declared that they had ''been troubled for years about indications that VDTs *may* cause health problems—and equally troubled by the fact that the press as a whole seems unwilling to report in any depth on the nagging questions about safety.'' They went on to say that ''so long as the major news media continue to practice a policy of benign neglect in regard to this story, we feel obliged to do what we can to fill the information gap,'' and they concluded by observing, ''It seems to us that there's a story here, even though it's not the kind of news that papers regard as fit to print.''

# THE SPANISH CONNECTION

As THINGS TURNED OUT, the call that Karel Marha made in the autumn of 1982 for experimental studies to determine whether pulsed low-frequency electromagnetic fields could produce harmful biological effects had, unknown to him, been partly answered. Five months earlier, Dr. José M. R. Delgado, a neurophysiologist who was director of research at the Centro Ramón y Cajal Hospital, in Madrid, had published the results of a discovery that would raise serious new questions about the effect of VDT radiation upon embryonic development. Dr. Delgado, who was then in his late sixties, had received his medical degree from the University of Madrid in 1940, and for the next six years had worked as an assistant professor of physiology at the university's medical school. In 1950, he came to the United States, and between then and 1974 he was a professor of physiology at the Yale School of Medicine, in New Haven, where he developed techniques for the electrical and chemical stimulation of the brain which he applied to studies of primate and human behavior.

A man with a flair for drama, Delgado conducted a series of highly significant, if somewhat controversial, experiments during his years at Yale. In 1954, he demonstrated that real pain can be produced by electrical stimulation of the brain. Subsequently, by electrically stimulating different spots in the brains of monkeys, he was able to make the animals behave like puppets. In a 1969 book, *Physical Control of the Mind: Toward a Psychocivilized Society,* he described how a human patient into whose brain an electrode had been implanted was forced against his will to make a clenched fist each time an electrical current was passed through the electrode. In a flamboyant exploit that made the front page of *The New York Times* in May 1965, he used one hand to wave a red cape before a bull in the bullring at Córdoba, and then stopped the enraged animal in mid-charge simply by pressing the button of a small radio transmitter he held in his other hand. (He had previously implanted an electrode in the subcortical region of the bull's brain, which controls all motor activity.) Not surprisingly, it was an open secret that much of Delgado's behavioral research at Yale was indirectly financed by the CIA.

During 1980 and 1981, Delgado and Jocelyne Leal, a cell biologist who led one of Delgado's research groups at the Centro Ramón y Cajal Hospital, set out to investigate the biological effects of weak ELF magnetic fields on the embryonic development of chicks. (According to Leal, they did not know at the time that such fields are given off by video display terminals.) In their experiment, they kept fertilized eggs from white leghorn hens in an incubator for 48 hours while exposing them to ELF magnetic fields of 10, 100, and 1,000 hertz, at intensities of 1.2, 12, and 120 milligauss. The results of this experiment, which were published in the May 1982 issue of the *Journal of Anatomy*—a highly respected, peer-reviewed British medical publication—were extraordinary and surprising. Delgado and Leal reported that 100-hertz magnetic fields of 12 milligauss in intensity had "a powerful effect on chicken embryogenesis, delaying or arresting it at a very early stage and limiting development to the formation of the three primitive layers, without signs of neural tube, brain, vesicles, auditory pit, foregut, heart, vessels, or somites." Indeed, fully 80 percent of the eggs they used in their experiment developed abnormally. They noted that malformations of the cephalic nervous system were particularly prevalent. They emphasized that they had used very-low-intensity pulsed magnetic fields, and that "thermal effects were therefore minimal and may be disregarded." They also pointed out that they had discovered power and frequency windows similar to those described by Ross Adey and Carl Blackman in their experiments with chick-brain tissue. For example, they found that magnetic fields of 100 hertz had a much greater inhibitory effect upon the development of chick embryos than fields of 10 and 1,000 hertz, and that a power intensity of only 12 milligauss produced a greater inhibitory effect than a power intensity 100 times as strong.

Because the *Journal of Anatomy* was not widely read by the American scientists who were engaged in investigating the biological effects of radio-frequency and other non-ionizing radiation, nearly a year passed before the startling findings of Delgado and Leal came to the attention of researchers in the United States. In the March 1983 issue of *Microwave News,* Slesin reported some interesting reaction to the findings of the Spanish researchers on the part of their American colleagues. Lionel F. Jaffee, a biologist at Purdue University, expressed skepticism because the magnetic fields were so weak. "I don't believe it, I can't believe it," he said. Carl Blackman, however, said that very-low-level effects were entirely possible, pointing out the extreme sensitivity of newts and salamanders to magnetic fields. Richard Tell, of the EPA's Office of Radiation Programs, declared that Delgado's findings were "amazing," and wondered why more scientists were not attempting to replicate his experiment. He added that if Delgado's results were confirmed, their significance could be enormous, with implications for a host of different technologies ranging from nuclear magnetic resonance imagers to light dimmers.

Later in 1983, Delgado, Leal, and a graduate student named Alejandro Ubeda published a follow-up paper in the *Journal of Anatomy,* describing new experiments which demonstrated that the shape of the magnetic-field pulse and the duration of its rise time—the time it takes to reach its peak intensity—might prove to be the "decisive parameter" in the inhibition of development in chicken embryos. Apparently, neither these nor Delgado's earlier findings were of sufficient magnitude to impress scientists at Yale. On January 4, 1984, Robert Handschumacher, of the Yale School of Medicine, told a special panel of Connecticut state legislators investigating the VDT health issue that existing radiation data from VDTs "do not warrant very much more attention." Robert Wheeler, a professor of applied physics at Yale, maintained that electromagnetic emissions from VDTs could be virtually eliminated by metallic shielding inside VDT casings.

(Wheeler must have forgotten that the metallic shields of VDTs are virtually transparent to the pulsed magnetic fields emitted by the machines.) At the same meeting, Dr. Florence Hazeltine, a specialist in obstetrics at the Yale School of Medicine, assured the legislators that the "apparent clusters" of fetal abnormalities and miscarriages among VDT operators were statistically insignificant. She added: "We do not advise women to stop working on VDTs while they are pregnant."

Even as Dr. Hazeltine spoke, NIOSH was investigating another cluster of pregnancy problems among VDT workers. This one involved VDT operators working on the fifth floor of the Southern Bell Telephone & Telegraph Company's data-processing center in Atlanta, where six out of fifteen pregnancies had ended in miscarriage—an event that the NIOSH official in charge of the investigation suggested was another "random occurrence." Still another cluster came to light on February 16, when an organization called 9 to 5, the National Association of Working Women, reported that between 1979 and 1984, twenty-four out of forty-eight pregnancies among VDT operators at the United Airlines reservation center in San Francisco resulted in miscarriages, birth defects, neonatal deaths, premature births, and other abnormal outcomes. According to the Air Transport Association of America, a trade group that was engaged in trying to prevent the enactment of VDT regulations and health measures by state legislatures around the nation, between 75,000 and 100,000 terminals —most of them operated by reservations clerks—were then being used by air carriers in the United States.

In the spring of 1984, yet another cluster—the eleventh to be reported since 1980—came to public attention when it was learned that seventeen out of thirty-two pregnancies occurring between December 1981 and March 1983 among VDT operators working in the Alma, Michigan, office of the General Telephone Company of Michigan, a subsidiary of GTE, had ended in miscarriages, birth defects, or other abnormal circumstances. NIOSH officials, who had long been discounting any link between VDT use and miscarriages and birth defects, now began to change their tune. On May 15, they announced that the long-delayed epidemiological study of reproductive risks among VDT operators had suddenly become a "very high priority," and that Dr. J. Donald Millar, the director of NIOSH, would describe the planned investigation later in the month, when he was scheduled to testify before the House Education and Labor Committee's Subcommittee on Health and Safety. At about the same time, a spokesman for the American College of Obstetricians and Gynecologists said that he would tell the subcommittee that radiation emissions from VDTs could not have caused the eleven reported clusters of miscarriages, birth defects, and other pregnancy problems.

Later that spring, *Microwave News* reported that Kjell Hansson Mild, of the Swedish National Board of Occupational Safety and Health, had confirmed the findings of Delgado and Leal, but that Mild had observed a 30 percent rate of malformation among chicken embryos, as contrasted with the nearly 80 percent rate seen by the Spanish researchers. By this time, the U.S. Office of Naval Research had become interested in Delgado's findings, and had made plans to send two scientists from the EPA to visit Delgado's laboratory in Madrid, and to fly two of Delgado's associates to the United States to meet with American researchers. Thomas C. Rozzell, a radiation specialist with the Office of Naval Research, had already visited Delgado's lab and was said to have been impressed with what he had seen there. According to a colleague, Rozzell had been told by the Spanish researchers that the orientation of the exposed chicken eggs in the earth's magnetic field could be a crucial factor in the experiments.

# THE TWISTING STAGE

SHORTLY AFTER THE CANADIAN CENTRE for Occupational Health and Safety issued its January 1983 press release calling for the shielding of computer terminals, representatives of IBM—the world's leading computer manufacturer—complained to officials at the Centre that their machines were already shielded, and that the Centre's recommendation for further shielding was unwarranted. IBM had, indeed, shielded its terminals, but not to prevent human exposure to electromagnetic radiation. It had shielded them to prevent electromagnetic interference (EMI) with other sensitive electronic equipment, and, more important, to guard against the possibility that leaks from terminals used for the storage and retrieval of classified information might be detected and decoded by an enemy engaged in electronic surveillance. In fact, the testing and shielding of such terminals had been conducted since the 1960s under a highly classified operation called Project Tempest.

IBM, it turned out, was a powerful supporter of the Coalition for Workplace Technology, which had been formed in 1983 by the Computer and Business Equipment Manufacturers Association—an organization representing more than 40 major firms—to

271

lobby against legislation designed to protect the health of VDT workers. By the autumn of 1984, the Coalition had signed up no fewer than 22 national trade associations in its crusade against VDT regulation, and had managed to block, table, or defeat most of the VDT bills that had been introduced in state legislatures. Some of the language used in the Coalition's campaign resembled that once used by the Air Force to persuade residents of Cape Cod that the pulsed radiation from PAVE PAWS was harmless. Charlotte Le Gates, director of communication for the Computer and Business Equipment Manufacturers Association, declared that for pregnant operators to ask to be transferred away from VDTs "is like asking to be transferred away from a light bulb."

As for IBM, it had for some months been participating directly in efforts to prevent VDT measures from being enacted by state legislatures. In January 1984, a company spokesman told Connecticut legislators that it would be premature to enact mandatory standards for VDTs. He called for more research on equipment and furniture design, and for ongoing public education about VDTs, and he recommended that studies of VDTs and workers be carried out in the "broader scope" of office automation. At about this time, IBM issued a booklet about VDTs which quoted a 1983 Canadian Radiation Protection Bureau study that said, "There is no reason for any person, male or female, young or old, pregnant or not, to be concerned about radiation health effects from VDTs." A few months later, the company opposed the decision of lawmakers in Rhode Island to set up a special panel to review VDT health risks.

Early in 1984, IBM had hired Professor Arthur Guy, the director of the Bioelectromagnetics Research Laboratory at the University of Washington, to analyze the literature on VDTs and measure the VLF electromagnetic fields emitted by VDT flyback transformers. Guy reported on some of his findings at a conference on "Office Hazards: Awareness and Control," sponsored by the University of Washington's Department of Environmental Health and held in Seattle in the late spring of 1984. Magnetic-field pulses from VDTs, he pointed out, were "characterized by a sawtooth-wave shape." (This shape results from a long magnetic-field pulse that regulates the horizontal tracing of a raster line from left to right by the electron beam of a VDT, and a very short electric-field pulse that causes the electron beam to fly rapidly back to the left-hand side of the screen to start a new line.)

Guy also reported that average magnetic-field strengths at the surface of the terminals were on the order of 10 milligauss, decreasing to between 0.2 and 1.4 of a milligauss at a distance of 10 inches or so from the VDT screen. He told *Microwave News* that the intensity of the current induced by the VDT magnetic fields he had measured was "about the same" as that induced in the chicken eggs in Delgado's experiments, and that the pulse shape of the VDT magnetic field was somewhat similar to a signal that was being used to heal bone fractures.

In the September 1984 issue of *Microwave News,* Slesin reported that an IBM spokeswoman had refused to tell him whether company officials had seen Guy's data on the VDT magnetic-field waveforms. According to Slesin, the spokeswoman said that a report that Guy had submitted to the company had been returned for revisions, and that IBM was not planning to make the final report public. At the end of October, however, IBM sent *VDT News*—a monthly newsletter Slesin had begun publishing earlier that year—a six-page summary of Guy's report, "Health Hazards of Radio Frequency Electromagnetic Fields Emitted by Video Display Terminals." The summary said that VDTs give off ELF fields of 60 hertz and pulsed VLF fields of between 15,000 and 25,000 hertz, but that the VLF fields were the more significant in terms of strength. It went on to say that VDT health effects had become a highly publicized issue when clusters of birth defects and spontaneous abortions were reported among VDT operators, but that "most investigators analyzing these problems have concluded that no causal link between VDTs and the reported effects can be established." As for the apparent similarities between VDT magnetic fields and the weak magnetic fields that Delgado and Leal had used in their experiments, the report had this to say:

Delgado and his associates have found that pulsed magnetic fields of certain waveshapes can produce teratological effects in chicken embryos. Though some investigators placed importance on these findings in relation to possible effects from VDTs, there are several reasons to discount the likelihood of such a relationship. Delgado's group saw the effects only with unique combinations of magnetic field waveform amplitude, rise time and shape. None of the combinations used for producing the effects had characteristics that simultaneously matched the VDT magnetic field waveform. The only characteristics in common with the VDT waveform were the amplitude and the rise time, but Delgado's group also observed nonsignificant effects for waveshapes with these same amplitudes and rise times, tending to rule out the significance of the parameters by themselves. Furthermore, the waveform used by the Delgado group had a recurrence or fundamental frequency more than 150 times lower and a pulse width 10 times greater than the VDT waveform. However, the implications of the Delgado group's findings in terms of NEMF [non-ionizing electromagnetic field] bioeffects demands careful replication of the work and further study of the phenomenon.

The summary of Guy's report concluded by declaring that "the author has found no valid evidence that would indicate any health hazard associated with NEMF exposures of persons operating VDTs."

Guy's claim that the waveform of the magnetic-field pulses used by Delgado and Leal did not exactly match those of the magnetic-field pulses emitted by VDTs was technically correct. However, considering the fact that IBM and other VDT manufacturers had long maintained such weak fields could not possibly produce harmful biological effects, this was hardly cause for comfort. Moreover, two weeks before IBM released its summary of Guy's report, Jane M. Clemmensen, a research engineer at SRI International's radio physics laboratory, in Menlo Park, California, had sent *Microwave News* a letter stating the importance of measurements not reported in the summary. Clemmensen noted that readings of pulsed magnetic-field emissions from VDT flyback transformers which she and some colleagues at SRI had taken were consistent with those reported by Guy. She went on to say, however, that "the electric and magnetic fields originating from the flyback transformers of VDTs are also accompanied by strong electric and magnetic fields at harmonics of the 60 Hz power line frequency," and that because the magnitude of these fields was approximately equal to that of the fields from the flyback transformer, these 60-hertz fields "must be measured in addition to VLF radiation."

In the spring of 1985, someone sent Slesin the full 66-page report that Guy had submitted to IBM the previous October. In it Guy had disclosed that some older VDTs emitted levels of radiation that might be considered to cause biological effects, and had recommended that those terminals be shielded. In the April issues of *Microwave News* and *VDT News* Slesin quoted Guy's recommendations in their entirety:

> Though it is highly unlikely that there is any relationship between the birth defect clusters and VDT emissions, the clinical work on magnetic bone growth stimulators and the magnetic field work of Ubeda et al. (1983) replicated by Mild (1984) does indicate that there could be a relationship. This perceived relationship prevails even though the waveform of VDT emissions differs markedly from those of bone growth stimulators and the Delgado apparatus. A major question, however, is the validity of the bone growth stimulation work and the reported effects by Ubeda et al. Until this validity issue is resolved, critics will use the results of the above works to argue that the level of emissions from VDTs are not safe. The localized E [electric] fields at the surface of an unshielded cover of a VDT nearest the flyback transformer can reach extremely high values as a result of the associated high voltage and close proximity of the transformer to the cover. Since these fields have a capability of inducing much greater currents in an exposed user of the device than the relatively low magnetic field emissions it certainly is desirable to shield the cover of the VDT. Since such shielding is relatively inexpensive the benefit to cost ratio is large. Such shielding is generally present in newer models of VDTs to satisfy FCC [Federal Communications Commission] requirements for reducing electromagnetic interference. Since the magnetic fields emitted by the VDTs are significantly lower and induce much less current in an exposed subject, there is less need to provide magnetic shielding. Therefore, unless it can be shown that there is a real hazard due to the magnetic field exposure such cost may not be warranted. Since the Ubeda et al. (1982) work implies that there may be a hazard, however, the work should be replicated and the data carefully analysed to determine whether further research is needed to answer the questions concerning the applicability of the results to the VDT magnetic field waveform. Such work should be carried out by a team of highly reputable teratologists and engineers to minimize or eliminate possible artifacts in the exposure systems and the biological assay protocols. Also a careful and thorough characterization of the induced fields and currents in subjects exposed to VDTs should be carried out and the levels compared to levels known to be safe based on the most reputable scientific literature.

In both content and tone, this paragraph was a far cry from the final sentence of the six-page summary, according to which Guy declared that he had found no valid evidence of health effects associated with magnetic-field emissions from VDTs. The discrepancy received further attention on April 16, 1985, when *The Wall Street Journal* ran a front-page item describing Slesin's disclosure that Guy had recommended shielding older VDTs in his full report. According to the *Journal* story, an IBM spokesman said that Guy did not "recommend" shielding, but only said that it was desirable. The *Journal* noted that Guy was unavailable for comment.

On April 30, IBM's director of information, Robert J. Siegel, wrote the following letter to *VDT News:*

> Your recent . . . story misinterprets a report prepared for IBM by Dr. A. W. Guy. Your story is inaccurate and grossly misleading.
>
> This is especially distressing since IBM provided you in advance with additional clarifying information. You also had Dr. Guy's report summary which states there is "no valid evidence that would indicate any health hazards associated with . . . persons operating VDTs."
>
> Dr. Guy since has added a preface to his full report which states "VDTs are safe to use." He adds that the emissions from "both older and newer model VDTs are well below the levels shown to cause harmful biological effects."
>
> Here, again, are the facts:
>
> —No place in the report or summary does Dr. Guy recommend refitting older VDTs with shielding because of possible health problems. Dr. Guy reconfirms this in his preface by saying: "I do not feel . . . that unshielded VDT emission levels represent a potential health hazard."
> —Dr. Guy does observe that there is a *perceived* problem based on *unverified* reports. In his preface, Dr. Guy explains that current shielding is desirable because it reassures individuals concerned by those reports, not because he has any health or safety concerns.
> —IBM has not kept Dr. Guy's findings confidential. In accordance with standard scientific practices, Dr. Guy has been reporting his findings at appropriate scientific forums and he will continue to do so. Also, Dr. Guy's summary has been available since September and the full report is available now.
> —Dr. Guy's basic conclusion is that VDTs are *safe* and present *no* health hazards to users.
>
> I have enclosed Dr. Guy's full report, including the preface.
> Because we believe that the public is best served by a discussion

of VDT safety based on the facts, I would appreciate your printing this letter in its entirety in your next issue.

Slesin obliged Siegel, by printing his letter in its entirety in the May–June 1985 issue of *VDT News* and in the June issue of *Microwave News*. Beside it, in both newsletters, he ran a reply of his own. This is how it read in *Microwave News:*

We strongly dispute IBM's contention that our story is inaccurate and misleading. We offered no judgment on the question of VDT safety. We simply reported what we consider to be a clear recommendation by Guy for shielding VDTs which do not meet the 1983 electromagnetic interference (EMI) rules set by the Federal Communications Commission.

Referring to older model VDTs in a section entitled "Recommendations" on page 56 of his report to IBM, Guy states: "The localized E-fields at the surface of an unshielded cover of a VDT nearest the flyback transformer can reach extremely high values as a result of the associated high voltage and close proximity of the transformer to the cover. Since these fields have a capability of inducing much greater currents in an exposed user of the device than the relatively low magnetic field emissions, *it certainly is desirable to shield the cover of the VDT"* (emphasis added). We regret that IBM does not address this recommendation in its letter.

Indeed, to avoid any possibility of misrepresenting Guy's advice to IBM, *Microwave News* reprinted the *full text* of his recommendations. We refer interested readers to page 11 of our April issue.

On receiving IBM's letter, we contacted Guy at the University of Washington in Seattle, but he refused to comment on either our story or Siegel's letter.

Guy's new preface, cited by IBM, is dated April 18; the original report is dated December 2. (By April 18, our April issue was already on its way to our subscribers.) While the preface may help everyone better understand Dr. Guy's position, in no way does it alter the contents of the report itself.

Last fall, when *Microwave News* first learned that Dr. Guy was preparing a report for IBM, we repeatedly asked for a copy. We were told that it was for internal use only. Later, IBM released a six-page report summary that did not include Guy's recommendations. Only after we published our story did IBM finally release the full report.

We stand by our story.

Slesin's refusal to be swayed by Siegel went unnoticed in the press. Attempts at persuasion on the part of IBM had, however, been reported in the past. "People will adapt nicely to office systems if their arms are broken," William F. Laughlin, an IBM vice president, told *Business Week* in 1975. "And we're in the twisting stage now."

Chapter

# IN FRONT OF EVERY NOSE

STARTING IN 1986, there was a surge in scientific activity regarding VDT radiation. Thanks largely to the efforts of Thomas Rozzell, of the Office of Naval Research, a major international project to replicate the findings of Delgado and Leal got under way when scientists at five different laboratories (ultimately, six laboratories participated) in the United States, Canada, Sweden, and Spain undertook to confirm the teratological phenomenon that had been observed by the Spanish researchers. The investigation, which became known as the Henhouse Project, called for the researchers in all six laboratories to use eggs from white leghorn hens and identical exposure equipment, and to follow the same carefully detailed protocol in conducting their experiments. It also called for them to conduct the experiments only between April and early June, and between September and October, in order to ensure that embryonic development would not be affected by unusual ambient temperatures. Since the experiments performed by Leal had indicated a key factor in the teratological effect was the orientation of the chicken embryos relative to both the earth's static magnetic field and the artificially pulsed magnetic field of the exposure system, it was decided that the project would investi-

gate the role of the earth's magnetic field. Among those participating in the study were Leal and Ubeda, of the Centro Ramón y
Cajal Hospital, in Madrid; Kjell Hansson Mild and Monica Sand-
strom, of the National Board of Occupational Safety and Health,
in Umea, Sweden; Alexander Martin, of the Department of Anatomy at the University of Western Ontario, in London, Ontario;
Jack Monahan and Arnold Fowler, of the Molecular Biology
Branch of the FDA's Center for Devices and Radiological Health,
in Rockville, Maryland; William Koch, of the University of North
Carolina, in Chapel Hill, who was collaborating with Ezra Berman, a veterinarian at the EPA's Health Effects Research Laboratory, in Research Triangle Park, and with William Joines, of
the Department of Electrical Engineering at Duke University, in
Durham; and Richard Tell, of the EPA's Office of Radiation
Programs, in Las Vegas, Nevada, who was assigned the task
of designing and building the identical magnetic-field exposure
systems.

IBM was not impressed with the Henhouse Project. Spokesman Thomas Mattia said that the company had "strong reservations concerning the reliability of Delgado's findings." Mattia
noted that egg studies had little or no relevance to mammalian
reproductive systems, and that there were "major and significant
differences between the frequencies and wave shapes used by
Delgado's apparatus and those generated by VDTs."

IBM was not alone in expressing reservations about new research on the biological effects of VDTs. In November 1985, the
BellSouth Telephone Company sent President Reagan's Office of
Management and Budget (OMB) a harsh critique of NIOSH's
planned epidemiological study, which by then had been designed
as a retrospective study comparing pregnancy risks among fifteen
hundred women who used VDTs regularly with those observed
among fifteen hundred women who did not use VDTs. The critique was written by Dr. Brian MacMahon, chairman of the Department of Epidemiology of Harvard University's School of
Public Health, and by Sally Zierler, an epidemiologist in the Department of Community Medicine of Brown University, who had
undertaken it as paid consultants of BellSouth. They claimed that
the NIOSH study design was critically flawed by inadequate def-
inition of pregnancy outcomes, incomplete medical records on
pregnancy outcomes, excessive data collection, lack of concern
for recall bias, and insufficient population size. In December,

OMB officials rejected the NIOSH protocol, ruling that it was poorly designed and acknowledging that the BellSouth critique had influenced their decision.

Even as BellSouth was raising objections to NIOSH's proposed epidemiological study, the results of completed experimental studies implicated VDT radiation more strongly as a cause of birth defects. On January 30, 1986, officials of the Swedish National Board of Occupational Safety and Health (NBOSH) held a press conference to make public the findings of a study conducted by Dr. Bernard Tribukait and Eva Cekan, a research teratologist, both of the Department of Medical Radiobiology at the world-renowned Karolinska Institute, in Stockholm, and Lars-Erik Paulsson, an engineer at the National Institute of Radio Protection. The three researchers had discovered that weak, pulsed magnetic fields similar to those emitted by VDTs—in this case, sawtooth-shaped pulses with field intensities of 10 and 150 milligauss—caused more congenital malformations in the fetuses of exposed mice than in the fetuses of unexposed control animals. At the press conference, Dr. Ricardo Edstrom, the Board's medical director, said that this result was "totally unexpected," and that "we can no longer rule out the possibility that radiation could affect fetuses." According to the press reports from Stockholm, Edstrom went on to say that the findings might force the Swedish government to change VDT work regulations to protect pregnant women.

In the United States, Tom Brokaw reported on the *NBC Nightly News* that a new Swedish study had shown that pulsed radiation from VDTs could cause fetal abnormalities in the offspring of pregnant mice. "The findings mean that we can no longer rule out the possibility that radiation could affect human fetuses," Brokaw said. As a result of NBC's report, the Swedish Embassy in Washington was swamped by inquiries. Two weeks later, however, officials of NBOSH issued a statement that tried to play down the results of the Karolinska study. "No hypothesis for an effect of the VDT magnetic field on humans has been confirmed," the statement said, "and those epidemiological studies that have been performed have not been able to demonstrate any effect as to reproductive outcomes."

Within a few days, Dr. Tribukait publicly rejected the methods that had been used to reanalyze the data collected by him and his colleagues, and revealed that he had opposed as premature the

National Board's decision to release the results of their study. Paulsson also said that the release had been a "mistake." A third scientist with knowledge of the situation described NBOSH's memorandum as "political business." Dr. Edstrom, who had been responsible for releasing the results of the Karolinska study, subsequently attacked journalists for misrepresenting his statements and causing anxiety among VDT operators. Commenting on this, *VDT News* noted that Edstrom "did not explain why it was okay to issue a press release but not for the press to report on it."

In the May–June 1986 issue of the *Columbia Journalism Review,* Slesin pointed out that in spite of Brokaw's mention of the Karolinska study on *NBC Nightly News,* no major daily newspaper in the United States had picked up the original Reuters dispatch from Stockholm, a circumstance "providing further evidence of a blind spot in print journalism" to the VDT radiation hazard. After describing the findings of Delgado and Leal, and the Henhouse Project that would soon get under way to verify them, Slesin criticized the extraordinary slowness with which NIOSH had proceeded to implement its four-year-old promise to conduct an epidemiological study of pregnancy problems among women who operate VDTs. He also criticized the Reagan administration's OMB for stalling the project. In addition, he deplored the fact that Dr. Irving J. Selikoff—professor emeritus of medicine at the Mount Sinai School of Medicine in New York City, and long acknowledged to be the world's foremost authority on the occurrence of asbestos disease—had been unable to find funding to perform a prospective study of VDT pregnancy problems. Slesin pointed out that "If the Delgado effect checks out, then other types of radiation—including radio and TV broadcast signals, radar, and power lines—might also be risky," and "Electronic pollution would then join toxic chemicals as one of the most intractable of our environmental problems." He concluded his piece as follows:

> If it turns out that VDT radiation is biologically active, the prevailing apathetic response to questions about VDT safety might well be regarded as irresponsible. People would no longer be asking why no one cares to know if VDTs can cause pregnancy problems, but why so few cared to investigate the issue. For journalists, there would

be no excuse, since the story had been there all along, right in front of every publisher's and editor's nose.

No leading newspaper in the United States sent a reporter to cover the first major international conference on VDTs, which was held in Stockholm in mid-May 1986. Among the researchers from thirty different nations who attended the conference was Lars-Erik Paulsson, who delivered a paper which disclosed that new data gathered by him, Tribukait, and Cekan continued to indicate a statistically significant increase in fetal malformations among the offspring of mice exposed to sawtooth-shaped magnetic-field pulses similar to those emitted by the flyback transformers of VDTs. The only American journalists at the conference were Slesin and Mark A. Pinsky, the editor of *VDT News,* who reported that two studies (one Swedish and one Finnish) showed increased risks of cardiovascular abnormalities in children born to women who used VDTs.

# IGNORING THE EVIDENCE

ON SEPTEMBER 2, 1986, Representative Ted Weiss, chairman of the Subcommittee on Intergovernmental Relations and Human Resources of the House Committee on Government, wrote to Otis Bowen, the Secretary of Health and Human Services, asking him to override a decision by the Office of Management and Budget to delete queries about stress and fertility from a questionnaire that NIOSH had proposed to send to some four thousand married women working at BellSouth and AT&T. In his letter Weiss pointed out that the congressional Office of Technology Assessment had warned that the OMB deletions would diminish chances that the results of the NIOSH study would be credible. On November 12, Bowen informed Weiss that he did not support NIOSH's objections to the OMB deletions, and that he believed the study questionnaire to be adequate.

Meanwhile, on October 2, Weiss wrote to James C. Miller III, the director of OMB, telling him that the subcommittee had concluded that OMB had exceeded its authority in restricting the scope of the proposed NIOSH study. Reminding Miller that the OMB's Office of Information and Regulatory Affairs had refused to approve the NIOSH study protocol after receiving complaints from BellSouth, Weiss suggested that BellSouth had a conflict of interest. "At the same time that BellSouth was opposing the questions on stress in the NIOSH VDT study, they were aware of a health hazard evaluation that NIOSH was conducting at BellSouth in North Carolina, to determine if VDT workers were suffering from angina," Weiss told Miller. "The evaluation, completed in May 1986, found that men and women who use VDTs are more likely to suffer from chest pain than workers who do not use VDTs. NIOSH also found that chest pain was more likely among workers who used VDTs more frequently, and who felt a lack of control in their jobs. Chest pain is usually associated with stress, as is a feeling of lack of control in one's job." Weiss also told Miller that failure to include the deleted questions—they included queries about birth control, smoking, and drinking—would weaken the NIOSH study, and that because some ten million Americans were using VDTs every day, research designed to assess potential VDT hazards was of great importance.

On November 5, Weiss sent Miller a follow-up letter, telling him that a recent study prepared by scientists from the Harvard School of Public Health and the Mount Sinai School of Medicine had found that "OMB has been significantly more likely to interfere with research in occupational health" than with other research proposals, and that "OMB clearance officers were especially unlikely to approve research on reproductive health hazards." He also told Miller that when BellSouth's consultant, Sally Zierler, criticized the NIOSH protocol, she didn't know that NIOSH was investigating the possible relationship between angina and VDT use among BellSouth employees. "As a result," Weiss noted, "Dr. Zierler now supports the inclusion of questions on stress and fertility." Weiss informed Miller that "in response to Dr. Zierler's concerns, BellSouth has apparently decided that it will no longer oppose the inclusion of stress and fertility questions." Since this was the case, Weiss went on, "we sincerely hope that OMB will renegotiate the agreement with NIOSH regarding the VDT study."

In a letter received on November 7, Miller replied to Weiss's letter of October 2, saying, "We required that NIOSH delete the questions about fertility because the study plan submitted to OMB contained no hypothesis relating fertility to VDT exposure and because the proposed study did not have the statistical power to test any such hypothesis." As for OMB's insistence that questions on stress also be deleted from the NIOSH proposal, Miller told Weiss that "if a relationship emerges in this study between exposure and miscarriages and if this relationship is confirmed by other ongoing studies, we believe that further research in this area is warranted to determine whether stress or radiation is the cause." (He seemed unaware that exposure to radiation can cause stress, and that the matter might not therefore be a question of either-or.) Miller predicted that the NIOSH study "will probably leave unresolved the issue of adverse health effects of VDT exposure," and he denied that OMB had been unduly influenced by BellSouth in making its decision to delete the question about stress.

In the end, the OMB refused to allow NIOSH to reinstate the questions about stress and fertility, thus reinforcing the likelihood that Miller's prophecy about its inconclusiveness would be fulfilled. When the study got under way in July 1987, Teresa Schnorr, a NIOSH epidemiologist who was in charge of the investigation, said that it would be completed in the fall of 1988. The study was delayed, however, and its results are not expected to be made public until the end of 1989—more than seven years after it was proposed.

While the Reagan administration was doing its best to sabotage the NIOSH study in the autumn of 1986, the epidemiological waters were further muddied when the *Journal of Occupational Medicine* published the results of an investigation that had been conducted by Dr. Alison D. McDonald, research director at the Institut de Récherche en Santé et en Sécurité du Travail du Quebec, in Montreal. Together with several colleagues—including her husband, Dr. J. Corbett McDonald, a professor at McGill University's School of Occupational Health, in Montreal— McDonald had gathered data on more than 100,000 current and previous pregnancies in eleven Montreal hospitals between 1982 and 1984, and had determined that 17,632 of the patients concerned had worked in occupations involving "substantial use" of VDTs. McDonald and her associates reported that in both the current and the previous pregnancies, the rate of children born with congenital abnormalities had been similar among VDT users and non-users, and that during the previous pregnancies there had been no difference in the rate of spontaneous abortion among VDT users and non-users. They found that in the current pregnancies there was a statistically significant excess of spontaneous abortions among some VDT users as compared with non-users—the largest increase being among women who worked at the machines for between seven and twenty-nine hours a week. They added, however, that this excess could have been the result of "biased recall" on the part of the women they had interviewed. As a result, McDonald and her associates concluded that their study "does not support the suggestion that work with a [VDT] in pregnancy increases the risk of congenital defect or spontaneous abortion."

Because McDonald and her co-workers had not designed their study as a broad general survey of pregnancy outcomes in Montreal, they had included only two questions about VDT work: whether a woman used a VDT and, if so, for how many hours a week. Moreover, by ascribing the excess of spontaneous abortions in recent pregnancies to faulty memory on the part of the women they had interviewed, McDonald and her colleagues laid themselves open to some heavy criticism. Jeanne M. Stellman, of the Women's Occupational Health Resource Center, in Brooklyn, New York, who is on record as being skeptical that VDTs can cause miscarriages, described McDonald's use of recall bias to explain a positive finding as "specious," pointing out that her reasoning "calls into question the validity of all census and socioeconomic data, which are based on the assumption that people can be relied upon to remember and report recent events."

Further indication that investigation into the association between VDT use and reproductive risks would continue to be controversial came in March 1987, when the Council on Scientific Affairs of the American Medical Association published a report in the *Journal of the American Medical Association* which concluded that "No association has been found thus far between radiation emissions from VDTs and reported spontaneous abortions, birth defects, cataracts, or other injuries." The AMA report went on to dismiss VDT radiation as a possible hazard, saying that radiation levels "have been well below presently accepted standards of exposure"—meaning, of course, standards designed only to protect against the heating of tissue.

As it happened, the AMA report excluded all of the papers that had been presented at the May 1986 international conference on VDTs in Stockholm, even though it included data that had been presented subsequently. Dr. William Hendee, the AMA's vice president for science and technology, told *VDT News* that it is "hard to know the significance" of the Stockholm papers until they have undergone peer review. He cautioned, however, that "We're not saying that the definitive evidence is in yet on VDTs."

In an editorial in the March–April 1987 issue of *VDT News*, Slesin pointed out that the AMA report had completely ignored the findings of the Karolinska Institute study, which he described as "incriminating," and called attention to the fact that no effort was under way in the United States to repeat this important investigation. A recent report on VDTs issued by the World Health Organization had also completely ignored the Karolinska study, even though two members of the panel that had written the report had attended the Stockholm conference. "Some readers may accuse us of putting too much attention on radiation effects," Slesin wrote. "Perhaps, but until someone goes to the trouble of showing that the Karolinska mice study is invalid, and until full-scale epidemiological studies show that VDTs are not associated with problem pregnancies, we will continue to maintain that there *is* a cause for concern, and not just about perceptions of risk."

Shortly after Slesin called for the Karolinska study to be repeated in the United States, Professor Gunnar Walinder, director of the Radiobiological Oncology Unit of the Swedish University of Agricultural Sciences, in Uppsala, and his colleague Hakon Frölen reported that they had found a significant increase in fetal deaths and fetal losses among pregnant mice exposed to weak pulsed magnetic fields compared with those occurring in non-exposed test animals. The Swedish researchers also found a higher (though not statistically significant) incidence of congenital malformations among the exposed animals. Dr. Tribukait told *Microwave News* that the difference in the rates of congenital malformation between the Uppsala study and the one he and his colleagues had performed at the Karolinska Institute might be due to the fact that the Uppsala researchers had exposed their animals for the first nineteen days of pregnancy, whereas he and his coworkers had exposed the mice in their experiment for only the first fourteen days of pregnancy. Tribukait suggested that "malformed animals might have been killed during the longer exposure period." In any case, he emphasized that the two studies had generally produced the same effect and that "it is rather clear that the effect is real."

In June 1987, Dr. Ingrid Nordenson, of the Department of Medical Genetics of the University of Umea, and Mild, of the Swedish National Board of Occupational Safety and Health, reported the preliminary results of an ongoing series of experiments: they found that radiation similar to that emitted by VDTs could cause genetic effects in exposed tissue samples. A salient aspect of the Karolinska, the Uppsala, and the Umea studies was that the radiation exposures in each of them had been designed to mimic as closely as possible the magnetic-field pulses that are emitted by VDTs.

Convinced that the latest results from Sweden were extremely important, Slesin issued a special *VDT News* press advisory about them on July 27, 1987, and sent it to twenty leading daily newspapers and magazines in the United States and Canada. He pointed out that the findings of the Uppsala study supported those of the Karolinska study. "These two sets of data, taken together, add credibility to claims that VDT radiation presents a risk to pregnant women," he noted. He also pointed out that although the VDT magnetic fields were weak, they were "of the same order" as the magnetic fields that had been linked to childhood cancer by the much discussed replication of Nancy Wertheimer's investigation which David Savitz had performed for the New York State Power Lines Project.

In the United States, the advisory was picked up by only one newspaper—New York *Newsday,* which on August 3 ran a small story about the new Swedish study on the third page of its business section. Ironically, just two and a half months earlier *Newsday* had run an editorial criticizing proposed legislation being considered by the Suffolk County Legislature to protect the health of VDT workers. The editorial had warned that a state or locality that discourages economically rewarding business from using computers "might pay a heavy price," and had gone on to resurrect the discredited industry bromide that radiation levels emitted by VDTs are "less than those produced by such common household appliances as irons and hair dryers."

Chapter

# TRAIL BLAZING

TOWARD THE END OF 1987, it emerged that Ontario Hydro and IBM would jointly sponsor a study of the possible link between pulsed magnetic fields from VDTs and adverse pregnancy outcomes. Declaring that "extensive studies to date show no proof of hazard," an IBM spokesperson nevertheless acknowledged that questions about VDTs and reproductive risks had been raised by the Spanish and Swedish research. Under the study protocol, four groups of mice, totaling about eight hundred animals in all, would be used in the experiment, which was to be conducted by scientists from the University of Toronto's Faculty of Medicine. According to Ontario Hydro's Stuart Harvey, who was designing the exposure system, three of the groups of mice would be exposed to 20,000-hertz VLF magnetic fields at three different intensities, and the fourth group would serve as non-exposed control animals. In an interview with *VDT News,* Harvey said that the exposure levels would range between 40 and 2,000 milligauss, much greater than the exposure of the average VDT operator, which he estimated to be about 2 milligauss.

The planned collaboration between IBM and Ontario Hydro

had some intriguing aspects. To begin with, IBM had been denying for at least seven years that there was any radiation hazard whatsoever from VDTs. In addition, the company had for at least four years engaged in numerous attempts to prevent VDT health measures from being enacted by state legislatures. For his part, Ontario Hydro's Harvey had declared in 1983 that VDT radiation posed no detrimental health problem. He arrived at this conclusion after conducting an experiment in which he measured only the VLF electric fields emitted by horizontal-deflection coils and the flyback transformers, while failing to measure either the VLF magnetic fields from the horizontal-deflection systems, or the 60-hertz electric and magnetic fields that were known to emanate from the vertical-deflection systems.

Subsequently, Ontario Hydro had engaged in a bitter dispute with the members of the Bridlehead Residents Hydro Line Committee, who, in the course of protesting the utility's plan to run twin 500,000-volt transmission lines within a few yards of a school in their community, had not only drawn attention to the findings of Wertheimer and Savitz concerning the increased risk of cancer in children exposed to power-line magnetic fields of only 2 milligauss, but had also furnished data showing that levels as high as 100 milligauss would be likely inside the school after the proposed lines were installed and activated.

The fact that the planned collaboration between Ontario Hydro and IBM contained no provision for investigating the biological effects of the 60-hertz magnetic fields that were known to be given off by VDTs took on additional significance when word got around about the results of a study that had been conducted by Jukka Juutilainen and Keijo Saali, of the Department of Environmental Hygiene of the University of Kuopio, in Kuopio, Finland, which were published in the *Scandinavian Journal of Work, Environment, and Health,* in 1986. The two Finnish researchers had measured ELF magnetic fields given off by seven different VDT models, and had found that in all cases the 50- or 60-hertz fields from their vertical-deflection coils were considerably stronger than the (20,000-hertz) VLF fields emitted by their horizontal-deflection coils. Considering that Wertheimer and Savitz had found an excess of cancer among children who had lived in homes where 60-hertz magnetic-field strengths were only two to three milligauss, the strength of some of the ELF magnetic fields emanating from the vertical-deflection systems of the VDTs was worrisome, to say the least. For example, Juutilainen and Saali measured a magnetic field of eight milligauss at a distance of about twelve inches from one of the machines, and a level of almost four milligauss at the operator position—about fifteen inches from the viewing screen—of another. They reported that these ELF magnetic fields dominated the 50- or 60-hertz power-line fields everywhere around a VDT except in the immediate vicinity of its power transformer.

Faced with what amounted to a virtual blackout in the American press of any reports about the Spanish and Swedish VDT research, Louis Slesin carried on during the winter of 1988 as best he could. In the January–February issue of *VDT News,* he published an editorial pointing out that the key conclusion to be drawn from the Karolinska and Uppsala studies was that pulsed magnetic fields similar to those emanating from VDTs could affect the fetuses of mice, and were thus biologically active. He noted that although it was clearly possible that VDT fields could affect the development of the human fetus, there were no federal health standards governing emissions from or exposures to the VLF or ELF radiation given off by the machines. He then put his finger on why and how this extraordinary situation had come about: "If it turns out that weak VDT fields present a risk to human health, it would force a downward revision for the health criteria of all types of non-ionizing radiation," he wrote. "This would undoubtedly cause severe impacts on a large number of key military and industrial interests. When seen in this context, the stakes in the outcome of the VDT issue are much higher than would at first appear."

A few weeks after Slesin's editorial, Ezra Berman, of the EPA, who had taken over from Rozzell as coordinator of the Henhouse Project, announced that the combined results of the six-lab chicken-embryo experiment indicated that extremely weak pulsed magnetic fields could indeed adversely affect the development of chicken embryos. Berman went on to say that researchers in five of the six laboratories had found increases in the number of abnormalities among exposed embryos, and that although the increase was statistically significant in only two of the labs, the data taken as a whole were highly significant. (The one lab that failed to produce the effect had been unable to secure the

white leghorn hens that were used by all the others and had sub-
stituted eggs from another strain of chicken.)

In a press advisory released on February 19, 1988, the Center
for Office Technology, as the Coalition for Workplace Technology was now called, tried to put the best face on these findings,
pointing out that the results of chick-embryo studies could not be
applied to humans with certainty. "Epidemiological data should
be given serious consideration when analyzing any finding from
animal research," the Center claimed. Once again, however, as
had happened so often in the nine years since VDT radiation had
first been suspected of causing miscarriages and birth defects, the
reality of biological research intruded upon the wishful thinking
of the computer manufacturers, who had seized upon just about
every possible pretext to deny that their machines could pose a
health hazard. In an article on "The Risk of Miscarriage and Birth
Defects Among Women Who Use Visual Display Terminals During Pregnancy" which appeared in the June 1988 issue of the
Mount Sinai School of Medicine's *American Journal of Industrial
Medicine,* Marilyn K. Goldhaber, Michael R. Polen, and Dr. Robert A. Hiatt, of the Division of Research at the Northern California Kaiser Permanente Medical Care Program, in Oakland,
California, reported the results of a case control study they had
conducted of 1,583 women who had attended Kaiser Permanente
obstetrics and gynecology clinics in the South Bay Area of San
Francisco during 1981 and 1982. It had been the purpose of an
initial, and much larger, investigation to study the effect of the
pesticide malathion on pregnancy outcome, malathion having
been sprayed in the area during 1981 and 1982 in order to combat
Mediterranean fruit flies that were threatening to ruin the state's
agricultural crops. However, in the autumn of 1982, prompted by
the many clusters of miscarriages and birth defects that were
being reported among VDT operators, NIOSH asked the researchers at Kaiser to include some questions on VDT use in
their study. When NIOSH and Kaiser subsequently could not
agree on how to collaborate, Kaiser went forward with the VDT
study on its own.

Goldhaber, Polen, and Hiatt wrote that they had found that
women who worked on VDTs for more than twenty hours a week
experienced a risk of both early and late miscarriage 80 percent
higher than the risk for women who performed similar work with-

out using VDTs. They also found that the offspring of the women who used VDTs had an increased rate of birth defects, although this finding was not considered to be statistically significant. Their article concluded:

> Our case-control study provides the first epidemiological evidence based on substantial numbers of pregnant VDT operators to suggest that high usage of VDTs may increase the risk of miscarriage. The implication of this finding is as yet unknown. No biological mechanism has been postulated, nor has a clear pattern of risk been observed across all occupational categories. Our data do suggest, however, the need for further investigations. Most needed are large cohort studies of working women that will provide objective measures of VDT exposure, ergonomic factors and job stress during pregnancy.

The Kaiser Permanente study finally started to shake the press out of its decade-long lethargy in reporting the adverse health effects of VDTs. In a two-column article in *The New York Times,* on June 5, 1988, Lawrence K. Altman, the newspaper's leading medical writer, described the results of the Kaiser study in considerable detail. Ten days later, the *Times* ran a front-page story by Philip M. Boffey, which began, "After years of relative quiet, the possible health effects of video display terminals have again become a public issue, fueled by a new scientific study and a new law on Long Island."

The law Boffey mentioned was a measure designed to protect VDT operators from eyestrain, stiff neck, and back problems, which had been passed by the Suffolk County Legislature on May 10. On May 26, the *Times* had published an editorial entitled "Suffolk's Reckless Screen Scream," which described the bill as "bizarre" and warned that it would set a "reckless precedent" for other legislators. "In a decade, video display terminals have revolutionized much of American life, including the newspaper business and this newspaper," the editorial said. "VDT's, with their television-style screens and typewriter-style keyboards, enable people to run mighty computers at work and publish volumes from home desktops. But there are fears of radiation dangers, or at least eye and back strain." The *Times* went on to claim that there was no basis for those fears, "or for anything more than continued watchfulness and the ordinary prudence associated with any other office or home machine, say an electric type-writer.'' The newspaper then castigated the Suffolk County leg-islators for requiring employers to pay 80 percent of the cost of annual eye examinations and new eyeglasses that might be needed as a result of VDT work. The *Times* did not acknowledge that a decade earlier it had been involved in a bitter dispute over workmen's compensation claims brought by the two copy editors who had developed cataracts after working on VDTs, or that during the previous six years it had not published a single word about the extraordinary Spanish and Swedish research on the harmful biological effects of VDT magnetic fields. ''It's one thing to blaze trails in coping with known environmental hazards like tobacco and pop bottles,'' the editorial concluded. ''It's quite another to set picky, commerce-inhibiting rules for hazards feared but not demonstrated. Such a measure may never be nec-essary. It is surely not needed now.''

---

# A WALK IN THE SUN

DURING THE LATE SPRING and early summer of 1988, a spate of newspaper stories and television reports about the results of the Kaiser Permanente study unsettled many people, especially pregnant women working with VDTs. Some health experts insisted that there was no need for pregnant VDT users to take any special precautions. Others advised women to transfer from VDTs while they were pregnant or attempting to conceive. Slesin suggested that they switch to laptop computers, which use liquid crystals instead of cathode ray tubes, pointing out that "women are most at risk in the early months of pregnancy, when they might not know they are pregnant." And Dr. Selikoff, of Mount Sinai, observed that an epidemiological study of the effects of VDT radiation on pregnant women "should have been done ten years ago."

Not all of the sudden new interest in the VDT health issue had to do with reproductive risks. Late in June, *The New York Times* ran a piece by Philip Boffey about the ocular hazards associated with the machines. Boffey began by saying, "Vast numbers of American workers are suffering eyestrain from prolonged work with video display terminals, most of it avoidable by wearing the proper glasses or changing the work environment, according to experts who have studied the problem." According to Boffey, the experts were of the opinion that working on VDTs would not cause permanent impairment of vision. Boffey did not, however, cite any specific research upon which this opinion was based. The closest he came to mentioning radiation was when he quoted Dr. Lowell Glatt, an optometrist in Hicksville, Long Island, who is a member of the American Optometric Association's study group on environmental and occupational vision. "The VDT is not a lethal weapon," Glatt said. "The VDT does not bombard the eye with mysterious things that are going to chew it up." He went on to tell Boffey that VDT work will not cause good vision to deteriorate, but "will take a weak link in the vision system, magnify that weakness, and take the system to a stress breakpoint earlier." According to Boffey, if VDT operators could change the position of their chairs, move their screens, and adjust the brightness and contrast of the images on the screens, their chances of incurring eye or muscular discomfort would be "greatly reduced."

Early in August, Boffey returned to the subject of VDTs and vision. This time, his article described the findings of a study that had recently been conducted by Dr. James Sheedy, an associate clinical professor of optometry at the School of Optometry of the University of California, Berkeley. Speaking at a conference sponsored by NIOSH, Dr. Sheedy reported that an evaluation made at the school's Video Display Terminal Eye Clinic of more than 150 patients who had worked with VDTs for an average of 6 hours a day over 4 years revealed that fully two thirds of them had difficulty focusing their eyes. "I do think we have to seriously consider whether the VDT might be causing some breakdown in the eye-focusing mechanism," Sheedy told Boffey.

At about this time, the Swedish Telecommunications Administration—Sweden's largest purchaser and user of VDTs—announced that it would retrofit, at a cost of about $500 apiece, all the machines that were being used by 3,500 directory assistance operators, in order to reduce their magnetic-field exposure. Here in the United States, four business firms on Long Island filed a lawsuit to overturn the VDT measure requiring eye examinations and workplace lighting improvements that had been passed by the Suffolk County Legislature. Gary Sazen, counsel for the Long Island Association—the region's largest business and civic group —described the reason for the lawsuit. "We want to give the

Suffolk County Legislature a message that they shouldn't tread in areas where they have no jurisdiction," he said.

Meanwhile, *The Washington Post* appeared to be having trouble understanding the findings and implications of the Kaiser Permanente study. On June 7, an article in the newspaper's Tuesday Health section reported that the study had found a "1.8 percent elevated risk of miscarriage in the first three months of pregnancy for women working at a computer terminal for more than 20 hours a week." The *Post* was off by a factor of a hundred. Actually, the study had determined such women were 1.8 *times*—or 180 percent—as likely to suffer miscarriages as non-users of VDTs. Two weeks later, the *Post* published a five-line correction. However, it soon committed another error. On August 20, *Editor & Publisher* printed an angry letter to the editor from Lawrence Wallace, the *Post*'s vice president for industrial relations, who charged that the headline of an article about the Kaiser Permanente study in the July 16 issue of *Editor & Publisher,* which read "Study Links Miscarriages to VDT Use," was "extremely misleading and inflammatory." According to Wallace, the authors of the study merely reported that women who worked more than twenty hours a week with a VDT had an "apparent" increased rate of miscarriage. "Life is difficult enough for working newspaper people, particularly females," Wallace declared. "Scare headlines add to the psychology and stress burden."

Wallace and the *Post* were soon set straight by Louis Slesin, in a letter that was published in *Editor & Publisher* on October 22. Slesin accurately quoted the authors of the Kaiser Permanente study as writing, "We found a significantly elevated risk of miscarriage for working women who reported using VDTs for more than 20 hours per week during the first trimester of pregnancy compared to other working women who reported not using VDTs." He went on to observe, "Instead of denying the risk—or engaging in wishful thinking—executives like Mr. Wallace should be encouraging the search for the cause of the increase and then finding solutions."

By this time, new data from Hakon Frölen, at the Swedish University of Agricultural Science, supported his earlier finding that pulsed magnetic fields could cause significant increases in fetal deaths and fetal losses among pregnant mice. "The fetus is most sensitive to [pulsed magnetic fields] in the early days of pregnancy," Frölen told *VDT News*. He added that he did not think pregnant VDT operators were at risk from VDT magnetic fields, because their exposure was limited, and because VDT fields were weaker than those he used in his experiment. Frölen's finding in mice was consistent with that of Jocelyn Leal, in Madrid, and of Dr. Alexander Martin, of the University of Western Ontario, both of whom reported that chick embryos were most sensitive to pulsed magnetic fields during the early stages of development. Meanwhile, Juutilainen and Saali, of the University of Kuopio, not only found this to be the case but also found that the head-tail axes of chick embryos have "a tendency to be oriented perpendicularly to the magnetic field" and "thus have properties resembling those of magnetic dipoles."

Except for the readers of *VDT News* and *Microwave News,* few researchers in the United States seemed to know about the disturbing Spanish, Swedish, and Finnish findings, and some researchers appeared not to care. On November 4, 1988, the National Institute of Child Health and Human Development (NICHD)—an agency of the National Institutes of Health—held a workshop meeting in Bethesda, Maryland, which was prompted in part by a proposal by Representative Weiss that instead of waiting for the results of the long-delayed NIOSH investigations, the NICHD should sponsor its own epidemiological study of reproductive risks from VDTs. Among those at the meeting who favored Weiss's proposal were Dr. Hiatt, one of the authors of the Kaiser Permanente study; David Savitz, who had confirmed Wertheimer's findings about the association of power-line magnetic fields with childhood cancer; and Teresa Schnorr, director of the ongoing NIOSH epidemiological study, who acknowledged that the question of whether VDTs caused reproductive risks could not be determined by her investigation alone.

Among those opposed to the National Institute of Child Health undertaking its own study was Dr. Joe Leigh Simpson, the chairman of the Department of Obstetrics and Gynecology of the University of Tennessee in Memphis, who was the chairman of the workshop. He told *Microwave News* that working at a VDT "is not any different from walking out in the sun." This analogy recalled statements by John Osepchuk, of Raytheon, who had once testified before Congress that the chances of being harmed by low-level non-ionizing radiation were similar to the chances of getting a skin tan from moonlight, and by Jerome Krupp, of the Air Force's School of Aerospace Medicine, who dismissed the hazard of the pulsed microwave radiation emitted by the PAVE PAWS radar on Cape Cod by declaring it to be "almost like sitting under a light bulb."

On December 9, *The Washington Post* ran a front-page story disclosing that since September 1987 fourteen miscarriages had occurred among women working on the fourteenth- and fifteenth-floor newsrooms at the headquarters of *USA Today,* in Arlington, Virginia. According to an informal survey taken in April of 1989, at least thirteen of thirty-six newsroom employees who had become pregnant after December of 1987 had suffered miscarriages. To begin with, there was speculation that the problem might be related to construction work taking place in the newsroom, or to the use of VDTs there. Later, it was suggested that excessive levels of lead in drinking water at the headquarters building might be to blame. A health hazard evaluation of the situation is being conducted by NIOSH.

# WATCHDOG

FOR NEARLY A DECADE during which most government officials, newspaper editors, and members of the medical and scientific community have avoided facing up to the serious potential health risks posed by electromagnetic fields from power lines and computer terminals, Louis Slesin has insisted upon the necessity of investigating the problem fully. A dark-eyed man of forty-two, he was born and brought up in New York City. He received a B.A. in chemistry from Johns Hopkins University, in 1968; an M.A. in chemical physics from Columbia University, in 1970; and a Ph.D. in environmental policy from the Massachusetts Institute of Technology, in 1978. In 1977, while working for the Natural Resources Defense Council (NRDC)—an environmental group based in New York and Washington, D.C.—he became interested in the biological effects of microwave and radio-frequency radiation, which was then just beginning to be recognized as a potential public health hazard.

During 1977 and 1978, Slesin researched the microwave problem for NRDC, hoping to encourage government and industry to take it seriously, and looking for ways to finance studies that would be necessary for promulgating exposure standards and health regulations. "It was a frustrating experience, because at that time you simply could not get a straight answer about even the most minor aspects of the problem," he recalls. "There was an extraordinary amount of obfuscation on the part of people in the military and the electronics industry, and a lot of wishful thinking on the part of practically everybody else that the whole issue just might get up and go away. Finally, it became clear to me that if I wanted to get accurate information about the microwave problem to the public, the best thing I could do was start up a newsletter."

Working out of his Manhattan apartment, Slesin published the first issue of *Microwave News* in January 1981. To begin with, the newsletter had fewer than one hundred subscribers; today, it has fewer than five hundred. Its influence, however, can be judged by the fact that it is read avidly by researchers in the field of non-ionizing radiation; by officials in some 30 governmental agencies, including the EPA, the FDA, the Departments of Defense and Energy, the FCC, and the FAA; and by officials of many large electronic and communication corporations, including IBM, RCA, GE, Hughes Aircraft, Raytheon, ITT, and AT&T.

Nevertheless, Slesin soon realized that *Microwave News* was not going to set the world on fire. "In November of 1981, Martha Zysko, an assistant who has been with me from the beginning, and I published a piece on the growing number of clusters of miscarriages and birth defects among VDT operators," he says. "I thought that our article would be snapped up by the newspapers and all hell would break loose. I couldn't have been more wrong. Few newspapers would even touch the subject, and there was scarcely any reaction. In 1982, we described some of the major lawsuits, past and present, that involved microwave radiation. There were about forty in all, and most of them had been settled out of court, with a gag provision imposed by the defendant companies on all information developed by the plaintiffs. This, of course, helped to keep a lid on the microwave problem, because it meant that each new plaintiff who came along would have to build his case from scratch. It also helped to convince me that a major cover-up was in progress."

Slesin is proud of the fact that in 1983, *Microwave News* became the first publication in the United States to take note of Delgado and Leal's research on weak pulsed magnetic fields. "A concerned official at one of the federal regulatory agencies tipped me off about their experiment," he says. "The thing to remember about it is that the strength of the fields that were affecting the chick embryos was minuscule. When I met Jocelyne Leal at a scientific meeting in Florence a year later, and told her that interest in her work centered largely on its possible relation to the magnetic fields being emitted by video display terminals, she was surprised because the VDT connection had never been brought to her attention. She was also surprised when I told her about all the clusters of miscarriages and birth defects that had been reported among VDT operators.

"Now that I think back on it, I find myself troubled that between 1981 and 1985, *Microwave News* and *VDT News* were just about the only publications in the United States that had any indepth reporting on those events. During that entire period, the people at NIOSH either didn't want to talk about the association between VDTs and reproductive risks or simply dismissed it out of hand. Except for Ross Adey, Robert Becker, Karel Marha, Carl Blackman, and a few other pioneers, most of the researchers in the field either believed or pretended to believe that non-ionizing radiation could cause injury only by heating up tissue. As a result, the problem of low-level, non-thermal effects remained shrouded in the realm of see-no-evil-hear-no-evil, while the computer manufacturers churned out VDTs by the million and put them in virtually every office, home, and school in the nation. The manufacturers were afraid that if it got out that the Delgado findings were important and relevant the whole electromagnetic universe would be suspect. That attitude also infected a lot of scientists and researchers who knew better but didn't speak out, because they feared that their research grants depended upon their continuing to give lip service to the obsolete notion that the only hazard from non-ionizing radiation was the thermal effects of high-intensity exposure."

According to Slesin, a major attempt to cover up the non-ionizing radiation hazard occurred in the summer of 1984, when the Air Force tried to claim that the results of the study performed by Professor Guy and his colleagues at the University of Washington on the effect of low-level microwave radiation upon rats were negative when in fact it was clear that the radiation had caused an extraordinary excess of cancers of the endocrine system among the exposed animals. "The important thing to remember about Guy's study was that it was the first and only experiment to test the effect of long-term subthermal doses of microwave radiation," Slesin points out. "If it had not been for Sam Milham speaking up at the Bioelectromagnetics Society meeting, the true story about its findings might never have got out, let alone made the *CBS Evening News.* That whole episode taught me never to underestimate the capacity of the military for deceiving the American people, and for being utterly brazen in their manner of doing so."

Slesin's next major revelation in *Microwave News* was the exclusive report he put out in March 1985 about Dr. Stanislaw Szmigielski's extraordinary epidemiological study, which had found that young Polish military personnel who were exposed to radar and other sources of microwave and radio-frequency radiation were seven times as likely to develop cancer of the blood-forming organs and lymphatic tissue as service personnel who were not exposed. "Szmigielski's study confirmed Guy's animal study with *human* data," Slesin points out. "But, incredibly, nothing happened. Not only were no epidemiological studies of American military personnel undertaken, but the Air Force was allowed to get away with not financing a replication of Guy's animal study. And, mind you, this went on even as we reported in *Microwave News,* in May of 1985, that all three radar repair technicians at the FAA station in Albuquerque, New Mexico, had either developed endocrine cancer or brain cancer. I mean, how much smoke does anyone have to see before figuring out that there may be a fire?"

Slesin says that a turning point in the long debate over the biological hazards of power-line radiation took place in the fall of 1986, in Toronto, when Richard Phillips, of the Environmental Protection Agency, declared publicly that he would not buy a home within a high-voltage power-line right-of-way. "That was a bombshell," he recalls. "No one ever dreamed that Phillips would say a thing like that. Practically everybody at that meeting was assuming that Savitz's replication of Wertheimer's study would not confirm her findings about the association between sixty-hertz magnetic fields and childhood cancer, and that the whole issue would be laid to rest forever. Yet here was a senior official of the EPA telling the meeting in the most vivid manner possible that he believed that the power-line hazard was for real. By the time Savitz announced, a month or so later, that he had in fact confirmed Wertheimer's findings, the sense of resignation in everybody—from the researchers in the field to officials of the utility industry—was almost palpable. In short, everybody knew that the problem was not going to go away."

Slesin points out that since then there has been a veritable cascade of publicity about the biological hazards of low-level magnetic fields, and that these revelations have put officials of the utilities and computer industries in a quandary. "If they acknowledge the validity of the Wertheimer, Savitz, and Delgado results, and all the other data, not only do they drastically increase the legal liability of their industries, but they also jeopardize their plans for future growth," he says. "So they have decided to keep on denying that a problem exists. The utilities people are being helped by the fact that few people seem to be aware that the magnetic fields associated with childhood cancer are emitted not just by the high-voltage wires that one sees strung across the countryside on huge steel and wooden towers but also by the ordinary high-current electrical distribution wires that are strung on telephone poles on the streets of practically every city, town, and village in America."

Slesin expects that electric utility officials and computer manufacturers will continue to deny the ubiquitous public health problem posed by electromagnetic radiation from power lines and video display terminals because their attorneys are advising them that this is the best course to follow. "It is a foolish tactic, however, because, just as toothpaste cannot be put back in its tube, the ELF- and VLF-radiation hazard can no longer be suppressed," he warns. "It is also irresponsible because the problem demands immediate action. Congress should initiate and finance a crash program of studies of the radiation hazard by independent agencies. The scientific investigation of this hazard should not be allowed to remain under the control of the military, the utilities, the Electric Power Research Institute, and the computer manufacturers, all of whom have a vested interest in its outcome."

# ASSESSMENT

THAT FURTHER INVESTIGATION of the ELF radiation hazard may lead to some highly discomforting findings can be seen from a recent experiment conducted by Reba Goodman, a geneticist and cell biologist at Columbia University's Medical Center, and Ann S. Henderson, a molecular biologist at Hunter College. Goodman, who received a Ph.D. in developmental genetics from Columbia in 1956, first went to work at the Medical Center in 1961. There she began to investigate transcription—how the genetic code is passed on from the DNA to the RNA, and thus from the nucleus of a cell into the cytoplasm, which is the region of the cell that lies between the nucleus and the outer membrane. "I investigated transcription by studying the salivary gland cells of the sciara fly, because they contain giant chromosomes that are highly visible and metabolically active," she explains. "What I was trying to find out is the secret of how the genetic code works. In other words, how a whole organism can develop from a single fertilized egg."

Goodman pursued this line of inquiry throughout the 1960s and 1970s—using various tools, such as hormones, to stimulate gene activity in sciara salivary cells—and she was still pursuing it in

1981, when she became aware that for some years she had been working in a laboratory that was on the same floor as a clinic that was being operated by Dr. C. Andrew L. Bassett, an orthopedist who had developed the use of electromagnetic signals to heal difficult bone fractures.

"I kept seeing people being rolled into Dr. Bassett's clinic in wheelchairs and later walking out under their own power," she remembers. "When I got around to asking how he was treating them and learned that he was stimulating bone growth with pulsed signals, I realized that he must also be stimulating gene activity. So, naturally, I got the idea that I ought to try stimulating gene activity in the salivary gland cells of my sciara flies with his bone-healing signals. Those signals, it turned out, were specially generated asymmetrical waves of fifteen and seventy-two hertz, which Dr. Bassett and his colleagues had designed to mimic electromagnetic ELF signals that are given off by broken and deformed bones. When I told Dr. Bassett what I wanted to do, he allowed me to borrow the generator and coil he used to produce those waves, and to use them in my experiment. In short order, I found out that when the salivary gland cells of sciara flies are stimulated with such signals, they produce more of a type of RNA that is known to carry the genetic code than unexposed cells do."

At that point Goodman realized that she did not have the necessary experience in molecular technology to analyze her observations, so she went across the street to get the advice of Ann Henderson, who had spent ten years at the Medical Center finding and mapping specific gene sites on human chromosomes. "She was interested in what I was doing, so we decided to investigate the phenomenon as a team," Goodman says. "In the course of subsequent experiments, we found that transcription could be induced after only fifteen minutes of exposure to the ELF signal. We also found that it could be detected both at the biochemical level, meaning all the different types of RNA that we could measure, and at the cytological level, meaning where the ELF-induced RNA chains were present on the DNA of a chromosome."

Between 1985 and 1987, Goodman and Henderson decided to test the effects of symmetrical signals of 60- and 72-hertz on genetic material, because these signals are similar to those of the electromagnetic fields that are carried by power lines. "We switched from using the salivary cells of sciara flies to those of drosophila, which are common fruit flies," Goodman says. "We studied the drosophila cells to see if their protein patterns could be induced or altered by exposing them to electromagnetic signals of five different frequencies, including the power-line frequency of sixty hertz. We found that all of the signals not only caused detectable changes in the protein patterns of the drosophila cells, but also increased their total protein production. This confirmed our previous observation that RNA transcription is extraordinarily sensitive to specific electromagnetic signals. This work was published in the *Proceedings of the National Academy of Sciences* in June of 1988. During 1987, we exposed cultured human leukemia cells to the five signals and measured common gene products, such as histone, actin, and *c-myc*. Histone is a protein that helps to regulate DNA. Actin is a protein that is found in many cells, and plays a role in cell division. *C-myc* is a protoöncogene—a normal protein component of all cells, but one that is also frequently found in tumor cells. During our experiments, we observed that each of these gene products was increased by exposure to the five ELF signals, but especially to the symmetrical signals. The fact that a sixty-hertz signal can enhance the expression of a specific oncogene may, of course, help us to understand how power-line radiation might act to trigger the development of cancer."

Together with the work of Ross Adey and his colleagues, the experiments of Goodman and Henderson provide additional evidence to support the association between power-line magnetic fields and childhood cancer that Nancy Wertheimer observed soon after she embarked upon her pioneering journey of discovery in the streets of Greater Denver fifteen years ago. Since then, power company officials, computer manufacturers, and some members of the scientific community, who are reluctant to admit the possibility that weak fields can produce biological effects, have greeted virtually every epidemiological study suggesting that 60-hertz magnetic fields cause cancer with the claim that the evidence was unsupported by information indicating the mechanism by which the disease could occur. A recent example of this took place on March 14, 1989, when the members of a seven-member panel convened by the National Academy of Sciences' National Research Council expressed doubt that weak ELF fields could play a role in the development of cancer. The panel members did so after hearing David Savitz describe the results of his childhood cancer study that confirmed Wertheimer's original findings, and of being warned by Savitz and Dr. David O. Carpenter, of the New York State Health Department, that the cancer risk of power-line radiation may have been underestimated as a result of inadequate knowledge on the part of scientists about the total exposure of the general population to magnetic fields.

One of the skeptics on the panel was Herman Schwan, professor emeritus of biomedical electronic engineering at the University of Pennsylvania and a long-time consultant to the electronics industry. Schwan developed the 36-year-old theory that the only way non-ionizing radiation can affect tissue is through heating; he testified at the New York Power Lines hearings that power-line fields were harmless; and he was instrumental in persuading the members of the National Academy's 1976 committee on the biological effects of ELF electromagnetic fields that such fields posed no threat whatsoever to human health.

Another of the skeptics was Robert Pound, a physicist from Harvard University, who in 1980 proposed heating people directly with microwave radiation in order to conserve energy in homes. Still another panel member, Dr. Clark Heath, of the American Cancer Society, described the epidemiological studies showing an association between magnetic fields and cancer as "interesting, but inconclusive," and the panel chairman, Richard Setlow, a biophysicist at the Brookhaven National Laboratory, on Long Island, who has specialized in the biological effects of ionizing radiation, described them as "interesting, but not convincing." What seems not only interesting but also alarming is that the childhood-cancer studies of Wertheimer and Savitz are accompanied by some twenty investigations from around the world showing that a significantly higher proportion of workers employed in electrical occupations—electricians, electrical engineers, utility-company servicemen—die of leukemia and brain cancer than do workers who are not occupationally exposed to electromagnetic fields.

Other scientists have expressed reservations about reports suggesting that the pulsed electromagnetic fields given off by computer terminals can cause birth defects and miscarriages. The fact remains, however, that the Kaiser Permanente study has shed disturbing light on the dozen or more unexplained clusters of birth defects and miscarriages that have occurred in North America during the past decade. So have the Swedish experiments showing that VDT pulsed magnetic fields can cause malformations in the fetuses of mice, and the Spanish studies showing that similar fields can adversely affect the development of chick embryos. Most unsettling of all, perhaps, is the fact that the pulsed VLF and ELF magnetic fields found routinely within a radius of about two feet from the average CRT computer terminal can be as strong as, or even stronger than, the sixty-hertz magnetic fields found inside the homes in which Wertheimer and Savitz discovered children to be dying unduly often of cancer.

As for the efforts of governmental regulatory agencies to investigate the health problem posed by low-level electric and magnetic fields, they have been inadequate. A case in point is illustrated by an article that appeared in the May 1989 issue of *FDA Consumer,* a magazine that the Food and Drug Administration publishes each month to provide information and guidance on health and safety issues for the public. The article starts out by claiming that the FDA has found "no conclusive evidence that electric blankets are a health hazard," and ends by declaring that the FDA "sees no reason for people to stop using electric blankets," but will "continue its research in this area and will monitor the research of other organizations." The second paragraph of the article reads:

> Electric blankets—like many other home appliances, including toasters, vacuum cleaners, and computers—produce low-intensity electric and magnetic fields. For the past decade, FDA and other scientists have been investigating whether magnetic fields affect human health. A study conducted in Colorado in 1979 suggested a relationship between exposure to electromagnetic fields and childhood cancer, but a similar study conducted in Rhode Island did not show such a relationship. And a 1988 study in Los Angeles County found no association between the use of electric blankets and adult leukemia. Animal studies have also proved inconclusive.

If the FDA were truly intent on keeping the public well informed about research on the biological effects of magnetic fields, why did it not point out that the average person is not chronically exposed over long periods of time to fields from most household appliances, like toasters and vacuum cleaners, but can be chronically exposed to fields emanating from power lines, electric blankets, and computer terminals? Why did the FDA not see fit to inform the readers of *FDA Consumer* that Wertheimer and Leeper's 1979 study of the association between childhood cancer and exposure to low-level magnetic fields had since been confirmed by the major epidemiological study that Savitz conducted for the New York State Power Lines Project? Why did the FDA fail to mention Wertheimer's 1986 study showing that the use of electric blankets and electrically heated water beds increases the risk of miscarriage? Also curious is the FDA's omission of the fact that a month after the *American Journal of Epidemiology* published the Rhode Island study, in its March 1980 issue, it published a letter from Wertheimer and Leeper, who pointed out that the Rhode Island researchers had collected birth addresses of their control population in a way that introduced serious bias into their study. In the letter Wertheimer and Leeper said that when they reworked the data to reduce the bias, they found that leukemia had, indeed, occurred more often among Rhode Island children living in homes near high-current wiring than among children living in homes near low-current wiring. What makes the FDA's omission even more egregious is that although the authors of the Rhode Island study were invited by the editors of the *Journal* to respond to Wertheimer and Leeper's critique, they declined to do so.

By expressing reservations about or by ignoring the epidemiological studies suggesting that low-level electric and magnetic fields may be hazardous to human health, and by questioning the experimental studies showing that ELF fields can alter brain chemistry in cats, produce learning disability in rats, and cause malformations in the fetuses of mice, the doubters within the government and the medical and scientific community have sought to discredit and disregard an entire body of medical and scientific data simply because it cannot be explained on the basis of existing theory. This does not provide a rationale for sound public-health policy, any more than the fact that the mechanism by which asbestos fibres cause cancer remains unknown could justify delaying the development of measures to prevent exposure to asbestos.

A far wiser policy with regard to the potential hazards of ELF and VLF electromagnetic fields would be to let the existing data speak for themselves—indeed, to err on the side of caution by assuming that they are valid—while undertaking a crash program of government-financed studies to determine the full extent of the risk that such fields pose to human health. The fact that scientists in the United States have yet to carry out a single experiment to determine either the short-term or the long-term biological effects of the pulsed ELF and VLF fields emanating from video-display terminals gives some indication of the research vacuum that exists. The fact that more than twenty epidemiological investigations already suggest that exposure to alternating-current magnetic fields from power lines and other sources is carcinogenic in human beings, together with the fact that pulsed electromagnetic fields such as those given off by computer terminals are known to be more biologically potent than non-pulsed fields, makes it imperative that these experiments be undertaken as soon as possible.

What may well delay the implementation of such badly needed research is fear among people in positions of authority and responsibility in the private sector, as well as in the state and federal governments, that if further investigation supports present indications that low-level electric and magnetic fields pose a health hazard, correcting the problem will prove to be not only tremendously expensive but also disruptive. It is undoubtedly for this reason that when officials of utility companies and state power commissions talk about the necessity of measuring magnetic fields from power lines, and about ways to reduce their strength, they are invariably talking about the magnetic fields given off by the high-voltage transmission lines that crisscross the countryside, and not those given off by the distribution wires feeding power into homes. For that matter, virtually all of the dozens of citizens' groups that have been formed to oppose the construction of power lines near schools and in residential neighborhoods are concerned only with high-voltage lines and not with local distribution wires. Yet it is precisely the ubiquitous neighborhood distribution wires, which Wertheimer and Leeper studied in Greater Denver during the mid-1970s, and Savitz and his colleagues restudied some ten years later, that are associated with the increased incidence of childhood cancer that both sets of investigators found in homes close to high-current wires.

A somewhat similar situation exists with regard to computer terminals, which are rapidly becoming ubiquitous. In 1976, there were fewer than a million VDT workstations in the United States; today, there are more than 30 million. According to the Computer and Business Equipment Manufacturers Association, almost one out of every fifteen white-collar workers now uses some type of computer workstation, and by the end of the century every white-collar worker will use one. Thus, if the magnetic fields given off by computer terminals should prove to cause cancer or otherwise be harmful to health, an immense and continually growing segment of the nation's population will have been placed at risk.

What can be done about a problem whose very magnitude inspires either silence or denial among most officials of industry and government, and among so many members of the medical and scientific community? The basic answer lies, of course, as it did in the recent case of the apple industry and the growth regulator Alar, with public awareness and public insistence that preventive measures be undertaken. The question is how to proceed. Some steps are more easily taken than others. Electric blankets, for example, can be discarded and replaced with blankets that have been redesigned to reduce magnetic-field emissions. Neighborhood power lines, however, pose a much more difficult problem. In some cases, restringing or burying them may be sufficient to reduce magnetic-field exposure. In other cases, it might be necessary to develop ways of changing the present grounding system, which, because it often allows unbalanced currents to flow through plumbing systems, can create relatively strong magnetic fields in many homes. The trick will be to accomplish this without losing protection against electric shock. Possible solutions include using nonconductive plumbing, or installing dielectric unions in existing plumbing to interrupt the flow of electric current in conducting pipe.

As for what to do about computer terminals, IBM and several other companies have provided an answer—albeit a somewhat selective one—by marketing in Scandinavia terminals shielded against VLF (though not ELF) magnetic-field emissions, in order to comply with Sweden's 1988 guideline specifying that VLF emissions not exceed the equivalent of one-half a milligauss at a distance of twenty inches. Although IBM has not introduced this model in the United States, a Norwegian firm is now marketing a line of VLF-shielded display terminals in this country. Meanwhile, a Massachusetts company has introduced a flat liquid-crystal display terminal that does not emit VLF magnetic fields, and has been shielded against ELF magnetic fields with a special metal called Mu-metal. An even more effective solution to the problem may be the one provided by the Fund for the City of New York—a nonprofit organization established by the Ford Foundation—which has designed its new offices so that all VDT operators sit at least 28 inches from their own terminals, and about 40 inches from other terminals. This, of course, is essentially the same solution that Karel Marha recommended almost seven years ago.

Thus, although the problem of electromagnetic-field exposure is potentially serious and obviously growing, there are a number of preventive measures to be explored, and some that can be implemented even as further studies are undertaken. Meanwhile, the de facto policy that power lines, electric blankets, and video-display terminals be considered innocent until proved guilty should be rejected out of hand by sensible people everywhere. To do otherwise is to accept a situation in which millions of human beings continue to be test animals in a long-term biological experiment whose consequences remain unknown.

# INDEX

## ABOUT THE AUTHOR

A longtime staff writer at *The New Yorker* magazine, Paul Brodeur is author of eight previous books, including *The Zapping of America* and *Outrageous Misconduct: The Asbestos Industry on Trial*. He lives on Cape Cod.

LaVergne, TN USA
27 February 2011
218042LV00002B/11/A